Cognitive Science Series

Series Editors:
Marc M. Sebrechts
Gerhard Fischer
Peter M. Fischer

Donald G. MacKay

The Organization of Perception and Action

A Theory for Language and Other Cognitive Skills

With 26 Figures

Springer-Verlag
New York Berlin Heidelberg
London Paris Tokyo

Donald G. MacKay
Professor of Psychology
Department of Psychology
University of California
Los Angeles, California 90024
USA

Library of Congress Cataloging-in-Publication Data
MacKay, Donald G.
The organization of perception and action.
(Cognitive science series)
Bibliography: p.
Includes indexes.
1. Learning, Psychology of. 2. Language acquisition.
I. Title. II. Series.
BF318.M32 1987 153 87-7995

Typeset by Publishers Service, Bozeman, Montana.
Printed and bound by Arcata Graphics/Halliday, West Hanover, Massachusetts.
Printed in the United States of America.

9 8 7 6 5 4 3 2 1

ISBN 0-387-96509-2 Springer-Verlag New York Berlin Heidelberg
ISBN 3-540-96509-2 Springer-Verlag Berlin Heidelberg New York

This book is dedicated to Deborah Burke
and Kennen Burke MacKay

skilled behavior, such as the mechanisms for sequencing, timing, and processing perceptual feedback in language and other cognitive skills.

I address this book to anyone with a serious interest in psychological theories, especially parallel distributed theories of relations between perception and action. However, I make no call in the book for converts to a new theoretical approach. Some readers may think that the field needs a different sort of theory, or that the predictions of my theory are uninteresting or unworthy of test. These readers nevertheless have something to gain from this book. Virtually every chapter begins with a set of requirements for a viable theory—the fundamental facts or constraints that any theory must explain. Although no current theory satisfies all of these constraints without modification, the constraints will remain as a challenge for anyone attempting to develop a more adequate theory in the future.

I have used this book as required reading for graduate and undergraduate seminars and as optional reading for large lecture classes in psycholinguistics, perception, and cognitive psychology. The 10 chapters of the book fit nicely into a 10-week quarter, and I have found that advanced undergraduates can readily comprehend all of the ideas developed here. I also hope that my ideas are comprehensible to anyone with a general interest in psychology, the brain sciences, and cognitive science disciplines such as artificial intelligence, philosophy, anthropology, linguistics, phonetics, and kinesiology. For these readers I have introduced technical language sparingly, I have defined generally accepted psychological terms, and I have described well-known phenomena and experiments in detail. Nonpsychologists can also take comfort in knowing that the bibliography at the end of the book contains references to background material.

I also address this book to colleagues and graduate students who are engaged in the research and simulation required for the development of theory in the cognitive and brain sciences. The book provides a detailed examination of topics long considered fundamental to researchers in psychology and related disciplines. I hope this second set of readers will forgive the descriptions of concepts and phenomena with which they are already familiar. In return they will discover in this book many new ideas for simulation and many new predictions for experimental test. Students of perception will find the discussion of consciousness in Chapter 4 especially provocative, and specialists in artificial intelligence may find Chapters 8 to 10 especially interesting from the standpoint of computer simulation. The relative noninteractiveness (locality or self-containedness) of the theoretical mechanism discussed there (self-inhibition) will tremendously simplify the task of simulating the theory's predictions and checking them against the available data. Specialists in the brain sciences should find Chapter 5 (The Temporal Organization of Perception and Action) especially rewarding; determining *which* brain events represent *which* processes is often quite difficult, but it is relatively easy to determine *when* a brain event occurs, which is the central theme of Chapter 5. Finally, empirically oriented psychologists will discover something of interest at many points in the book; there are new experimental findings, reported here for the first time, but obtained originally without the benefit of explicit theory. Although I say *new*, some of these data are now 20 years old, gathered at a time when my own orientation was primarily empirical. I had long ago filed these data away as either not making sense or not fitting the rapidly changing *Zeitgeist* in the field. What prompted me to dig out these "old" data was a new perspective on the vicissitudes of the *Zeitgeist* (D. G. MacKay, in preparation) and a new theory (developed in the remainder of this book) within which these data finally made sense.

I am pleased to record my indebtedness to the many colleagues and students who have commented on and thereby improved the contents of this book. I am especially grateful to Jay McClelland at Carnegie-Mellon University; Deborah Burke and Leah Light at the Claremont Colleges; Carol Fowler at Dartmouth College; Werner Deutsch, Uli Frauenfelder, Marie-Louise Kean, Pim Levelt, and Chris Sinha at the Max-Planck-Institut für Psycholinguistik, Nijmegen, The Netherlands; Wayne Wickelgren at the University of Oregon; Gary Dell at the University of Rochester; Andrew Comrey and Jerry Kissler at UCLA; and Alan Allport, Bruce Bridgeman, Lex van der Heiden, Herbert Heuer, Steve Keele, Dom Massaro, Doug Mewhort, Odmar Neumann, Wolfgang Prinz, and Eckart Scheerer, all members of the Research Group on Perception and Action at the Center for Interdisciplinary Research (ZiF) at the University of Bielefeld, Federal Republic of Germany. They have prevented many mistakes, and without their help, some of the conceptual butterflies in this book might still be caterpillars, but they are not responsible for any additional mistakes or infelicities that I may have made despite their advice.

I also thank John O'Connor at UCLA for the financial assistance and leave of absence during 1984 –1985 that enabled me to write the book, and I thank Herbert Heuer, Odmar Neumann, Wolfgang Prinz, and Peter Wolff for the invitation to join the Research Group on Perception and Action at the ZiF in Bielefeld, where much of the book was written. I also thank Wolfgang Prinz for the opportunity to help organize a conference entitled "Common Processes in Listening, Speaking, Reading, and Writing," which took place at the ZiF in July 1985. The many provocative ideas that arose from that conference (D. G. MacKay et al., 1987) provided a valuable stimulus for the present book. Finally, I thank the state of Nordrhein-Westphalia for providing the necessary funds and support facilities for carrying out these activities at the ZiF.

For their skillful typing of the first draft of the book, I thank Lorraine Cronshaw, Kathy Hacker, Charlana Watling, and the UCLA Central Word Processing Unit. For their help with word processing and computer programming, I thank Nancy Back and Lynn Thomas. For their help in running subjects and analyzing data, I thank Robert Bowman, Michael Birnbaum, and Brian Burke. For their help in proofreading the final copy and compiling the subject index, I thank Kent Bullard, Rana Matteson, and Julie Platus. Finally, I thank Pim Levelt and William Marslen-Wilson for providing two research fellowships at the Max-Planck-Institut für Psycholinguistik, where the book was completed.

Contents

Introduction

> The scientific description of verbal behavior (by linguists, of course, not by psychologists) is far advanced over any other area of behavioral description and so provides a glimpse of what other behavioral theories may look like eventually.
>
> (Miller, Gallanter, & Pribram, 1960, p. 154)

I wrote this book during a very enjoyable year that I spent in West Germany with a group of colleagues from universities around the world. All 20 of us were interested in the relations between perception and action, and together we formed the Research Group on Perception and Action organized by Wolfgang Prinz at the Center for Interdisciplinary Research (ZiF), located in the woods above the University of Bielefeld.

Before coming to Europe, I saw the year as an opportunity to broaden my interests. I already had formulated some ideas on the relation between speech perception and production, and the ZiF provided an interdisciplinary context that enabled me to examine these ideas from different perspectives. At the ZiF I had a chance to interact with colleagues from a variety of backgrounds – not just cognitive psychology, developmental psychology, and psycholinguistics, but neurobiology, neurophysiology, neurology, kinesiology, epistemology, and the philosophy of mind. The influence of these various perspectives can be found throughout the book.

My initial plan for the year was to determine whether my ideas on speech perception–production applied to other types of perception and action and to expand this comparison into a detailed and coherent book on the relations between perception and action. With this plan in mind, I wrote a preliminary paper comparing the organization of speech and visual perception, and I collaborated with John Annett (University of Warwick) on an experimental project exploring relations between speech and everyday actions such as tying one's shoelaces.

However, I soon found that my initial goal was too broad. The topics of perception and action have been of interest for so long and to so many disciplines, and have been approached with so many different methods, theoretical frameworks, and lexicons of description as to make detailed integration difficult. Even within psychology, so many findings are relevant that coherence requires selectivity.

The problem is that a coherent account can expect criticism for seeming to select arbitrarily from among the wide array of available data, and this led me to think about possible principles for data selection. The problem scarcely arises in advanced sciences such as physics and biology. The reason is simple. These fields have developed generally accepted theories that automatically provide the principles for selecting examples. By enabling physicists and biologists to see a wide range of facts as examples of the same underlying principle, theories have also reduced the complexity and conceptual heterogeneity of these fields. However, there exists no generally accepted theory in psychology, certainly not a theory capable of providing principled selection and simplification of available data on relations between perception and action. So, I decided to develop a theory of my own that synthesizes the wide range of perception–action issues that I myself have worked on over the past 25 years: the nature and causes of errors in the perception and production of speech and other cognitive skills, the mirror-image problems of ambiguity in perception and synonymy in production, the role of self-produced feedback in the integration of behavior, relations between timing and sequencing in perception and action, and relations between the cognitive and the physical aspects of action control.

I call my theory the "node structure" theory of perception and action, and I wish to point out some of its limitations at the outset. First, the theory only applies to the organization of *skilled* behavior, and many interesting topics fall outside its scope. For example, the topic of creativity necessarily involves behaviors that are unlearned, unpracticed, and therefore, unskilled. Nevertheless, I covertly took creativity into consideration in constructing the theory. Although a theory cannot attempt to explain everything at once, it must at least be consistent with established phenomena falling outside its limited scope. Awareness, attention, and learning are other important topics that I leave virtually untouched here but intend eventually to include in a more comprehensive theory (D. G. MacKay, 1987).

Another limitation of the theory concerns its emphasis on speech production–perception. Although I discuss other skilled behaviors such as piano playing, typing, and the generation of Morse code, I give the lion's share of attention to speech. Speech is more proficient (D. G. MacKay, 1981) and better described than other cognitive skills (Miller et al., 1960), but I believe that the principles of speech perception–production apply more generally to other perception–production systems, and I support this belief with enough examples throughout the book to justify including *Other Cognitive Skills* in the title. However, the issue of whether speech is unique promises to occupy the field for many decades to come, and I don't pretend that the present book has put it to rest. As Anderson (1980) notes, "The status of language is shaping up to be a major issue for cognitive psychology. The issue [of the uniqueness of language] will be resolved by empirical and theoretical efforts more detailed than those reviewed [here]" (p. 398).

The theory's essentially psychological focus represents another of its limitations. Like my colleagues at the Center for Interdisciplinary Research, I feel that interdisciplinary teamwork will eventually prove essential for establishing a theory of the relation between perception and action; and like McClelland et al. (1986), I feel that my theory can be readily mapped onto neuroanatomical constructs. However, the present book develops very few of these mappings. It touches only lightly on physiological data concerning effects of brain damage on sequencing and timing, perceiving and acting. Linguistic data on the structure of sentences have also fallen outside the scope of the theory. Unlike Chomsky (1957), I am not trying to describe the competence underlying our ability to produce all possible sentences in English or any other language. Nor do I attempt an exhaustive description of how we produce even a single sentence. All sentences are produced with some intonation, for example, but my theory includes no account of intonation. Of course, keeping the goals of physiology and linguistics in the background is not equivalent to disregarding them altogether. Both linguistic and physiological considerations will arise at various points, if only as reminders of phenomena that a more general theory eventually must explain.

Some might consider the relatively unformalized character of the theory a serious limitation. I am not so sure. Most scientific theories–for example, the wave theory of sound and the atomic theory of matter–began with qualitative descriptions before acquiring their more sophisticated mathematical form, and even with quantitative expressions firmly in place, a qualitative formulation invariably remains and provides the basis for using and extending a theory to new domains (Holland et al., 1986; Thagard & Holyoak, 1985). Moreover, as McClelland and Elman (1986, p. 13) point out, premature concern with formal or computational adequacy can obscure attention to fundamental properties of a theory such as its ability to account for available empirical laws.

I therefore view the qualitative state of my theory as a necessary precursor to later stages of development. The theory is in progress, a first step in the right direction. Three potentially parallel processes for developing the theory lie ahead: (1) real-time computer simulations resembling those of McClelland and Elman (1986) in some respects; (2) conceptual extensions (e.g., introduction into the theory of mechanisms for attention, awareness, and learning in response to novel experience; D. G. MacKay, 1987); and (3) experimental tests. When tested, some of the currently formulated predictions of the theory will turn out to be wrong, but the research required to simulate and test these predictions will provide the basis for a more complete and accurate theory in the future. Theories either stimulate advances in our understanding and knowledge or pass from the scene. However, this book will still be of value once the shortcomings of the present theoretical formulation have become apparent. Virtually every chapter spells out a set of fundamental facts or constraints that any viable theory must explain, and these constraints will remain as a challenge for anyone attempting to replace my own theory with a more adequate one in the future.

Comparisons With Other Theories

The node structure theory resembles other current theories in many respects. For example, like McClelland and Elman (1986) and Marslen-Wilson and Tyler (1980), the node structure theory deals with dynamic, on-line, or real-time perceptual processes. However, the node structure theory is much broader in scope than other current theories. It deals not just with perceptual processes but also with action and the relations between perception and action: for example, the units underlying perception and action, how they are activated in sequence, and how they are timed, not just how perception–production systems *construct* time—as happens when one produces a sentence at a voluntarily determined rate—but also how they *alter* time—as happens when one repeats the same sentence at different rates. The theory integrates a wide array of data on sequencing and timing, providing new explanations for phenomena such as constant relative timing, effects of practice on timing and sequencing in behavior, and regularities in the nature of sequential errors (e.g., speed–accuracy trade-off, the sequential class regularity, and the level-within-a-system effect (the fact that the probability of error is greater for lower than higher level units within the same system).

Unlike most other theories, the node structure theory postulates shared rather than completely separate units for perceiving and producing cognitive skills. These shared perception–production units provide new ways of conceptualizing well-known perceptual phenomena such as interactions between the timing of perception and action, perceptual–motor adaptation, categorical perception and its exceptions, contextual effects on phoneme perception, visual dominance and its exceptions, perceptual errors, and the time to comprehend ambiguous sentences. Shared perception–production units also provide the basis for some new hypotheses about the role of internal and external feedback in detecting and correcting errors, about differences in how we perceive self-produced versus other-produced speech, and about the cause of feedback-induced stuttering and its non-pathological analogue, the effects of auditory feedback that is amplified and delayed by about 0.2 s (Fairbanks & Guttman, 1958).

What is novel about the theoretical ideas I have developed is their integrative combination, and Chapter 1 begins by relating my theory to the 2000-year-old tradition of thought and experiments on relations between perception and action. Later chapters compare detailed aspects of the node structure theory with current theories of either perception or action or both. Theories discussed include McClelland and Elman's (1986) TRACE theory of speech perception; McClelland and Kawamoto's (1986) theory of ambiguity and "shades of meaning"; Marslen-Wilson and Tyler's (1980) cohort theory of word recognition; the motor theory of speech perception; Norman and Rumelhart's (1983) theory of sequencing and timing in typing; chain association and scanning theories of sequencing; Klapp, Anderson, and Berrian's (1973) buffer theory of production onset time; stage of processing theories of timing; Shapiro's (1977) programming theory of constant relative timing; efference copy/corollary discharge theories of how action influences perception; editor theories of error detection and correction; and the feedback control theory of Adams (1976) and others. Here I want to map out some more general relations between my theory and two broad classes of theoretical alternatives: production systems and parallel distributed processing (PDP) theories. I will argue that in general PDP models are too parallel and production systems are too serial to handle available data on relations between perception and action. The node structure theory represents an integration of these approaches that attempts to overcome the weaknesses of both.

Comparison With Parallel Distributed Processing Theories

Like the PDP theories of McClelland, Rumelhart, and the PDP Research Group (McClelland et al., 1986; Rumelhart, McClelland, & the PDP Research Group, 1986), the node structure theory represents knowledge in the connections between nodes and describes mental processes in terms of inhibitory and excitatory interactions occurring in parallel between nodes in a highly interconnected network. Another similarity is that nodes in the node structure theory simultaneously integrate many different sources of information so as to capture the multiple simultaneous constraints seen in behaviors such as word recognition, the motor control of typing, and the grasping of objects in everyday environments (Rumelhart, McClelland, & Hinton, 1986).

The theory also exhibits content addressability in its retrieval of information from memory, emergent properties such as constant relative timing and the processing of self-produced feedback, and graceful degradation—at least in the case of nodes representing the form or content of perception and action. Destroying a single content node will have only minor and difficult-to-detect effects on performance of the overall system, and as more and more content nodes are destroyed, performance will deteriorate gradually rather than catastrophically. Unlike serial symbol processing systems, disrupting a single step cannot incapacitate the entire system (Rumelhart, McClelland, & Hinton, 1986).

These similarities aside, my theory differs from PDP theories in several important respects. One is its distinction (discussed in detail in Chapter 3) between activation (a nonautomatic and sequential process that causes connected nodes to become primed and is necessary for conscious awareness and action) and priming (an automatic and parallel process that prepares a connected node for possible activation). Although other theorists have seen the need for such a distinction (e.g., Lashley, 1951; Mandler, 1985; McClelland & Elman, 1986, p. 77), the node structure theory is the only theory making systematic use of the priming-activation distinction.

Another difference is that the node structure theory represents "sequential rules" directly, rather than or in addition to indirectly, as properties arising automatically from the normal functioning of the network. This contrasts with McClelland and Elman (1986) and other PDP theories, where "linguistic rules" *only* receive indirect representation as emergent properties of the network. These sequential rules provide a natural description for sequential behaviors such as the rapidly produced sequences of phonemes in speech production, a problem for purely PDP models such as Boltzmann machines (Hinton & Sejnowski, 1986), which are notoriously bad at representing sequence.

A third and not altogether minor difference is that the node structure theory does not allow mutual inhibition between content nodes. Mutual inhibition between content nodes would make language production impossible in the node structure theory, despite its potential benefits for perception (McClelland & Elman, 1986). The node structure theory obtains these same benefits via inhibitory relations between *sequence* nodes, the control structures that activate content nodes during both perception and production.

A fourth difference concerns the flexibility of conscious versus unconscious processing. PDP theories provide viable accounts of subcognitive or unconscious processes using small (subsymbolic or microstructural) units and may eventually offer a description of cognitive or conscious processes as well (Rumelhart, Smolensky, McClelland, & Hinton, 1986). However, current PDP theories will have difficulty with the fact that normally unconscious processes can become conscious. For example, the microstructural becomes conscious and cognitive when we become aware of making a subphonemic slur during speech production (D. G. MacKay, 1987). Unlike PDP theories, the node structure theory provides mechanisms for representing the flexibility of conscious versus unconscious processing.

Finally, we come to an area where similarities and differences between the node structure theory and PDP theories form a complex mix: the issue of distributed versus central, localized, or local processes and representations. That is, discussions of distributed processing must distinguish between three different senses of the term *distributed*. One sense contrasts distributed with *central* processes, meaning roughly that distributed processes lack a central executive upon which all processing depends (as in production systems). In this sense, the node structure theory is definitely a distributed theory. A second sense concerns the issue of whether a processing characteristic is distributed throughout a system rather than localized at a particular processing stage in a sequence of stages for specifying the output. In this sense, too, timing, sequencing, and the form or content of perception and action depend on distributed processes and mechanisms in the node structure theory.

The third sense of distributed concerns *local* versus *mass action* representations of information for perception and action. Local theories represent a single concept with a single node (although they don't necessarily use a single node for representing only one concept; Chapter 2), whereas mass action theories represent a single concept with many nodes and use each node to represent many concepts. Information is represented not by particular units but by patterns of activity among large numbers of units. Some of the units in the PDP theories of McClelland et al. (1986) are not distributed processors in this sense and neither are the content nodes in the node structure theory. Each content node codes a given piece of information uniquely. Interestingly, however, timing in the node structure theory seems to be distributed in the mass action sense: The same timing node may represent different rates of output by varying its periodicity or rate of spontaneous reactivation, and a set of coupled timing nodes are required to generate any given rate of output.

Comparison With Standard Production Systems

Production systems (e.g., Anderson, 1976) represent knowledge via quasi-linguistic units and represent mental processes via sequential inferences resembling those observed during the conscious introspection of ongoing thought. More specifically, 'conditions' are cyclically matched against 'condition–action rules,' and a production becomes activated when its condition is met. Only one production can be activated at a time, and a central executive oversees the general flow of information and calls up subroutines to carry out a given task. As Hofstadter (1985) points out, the abstract and serial properties of production systems seem well suited for characterizing conscious but not unconscious, subcognitive, or microstructural processes.

Like production systems, the node structure theory represents rules directly, but its sequential rules are not the building blocks or basic units of the theory; they are called up in parallel and are triggered by a competitive activation mechanism rather than by a central algorithm for choosing which rule to trigger. Also unlike production systems, the node structure theory has no centralized processor for directing the flow of information and no condition–action rules for matching against conditions and triggering productions. Also, more than one sequential rule can fire at any given time or cycle, because sequence nodes in different "systems" can be activated simultaneously. Finally, the node structure theory combines parallel and sequential processes in a way that is not seen in production systems. For example, content nodes are activated in sequence but are primed in parallel.

Themes of the Book

Three major themes run throughout the book: representation (which units represent perception and action?), processing (what are the processes underlying perception and action and how do they differ?), and the integration of perception and action, as occurs during the perceptual processing of self-produced feedback. Readers interested in either representation, processing, or the integration of perception and action, but not in all three, are warned that these issues are not completely separable, and that the contents of this book are cumulative. Each chapter builds on information from the previous chapters, and reading the book from start to finish is advisable on the first pass.

Representational Issues

Representational issues in the present book focus on what I call the "middle ground" of cognitive psychology (see also Marr & Poggio, 1977; Rumelhart, McClelland, & Hinton, 1986). My strategy has been to start upstream at the highest level where conceptual waters seem clear and to work my way down as far as the available light permits. In the case of speech, the highest level for me has been the sentence and the lowest level has been the phonological feature. I found very little to say about the complex structures that intervene between sensory receptors (such as the basilar membrane) and the phonological features for speech perception, and I found even less to say about the structures that intervene between phonological features and the muscles for producing speech sounds. I do not claim these lower level perception–production structures are unimportant or that psychology should abandon its attempts to understand them. It is only that for molar units such as the sentence, structures at these levels are extremely complex and variable and in any case are beyond my current capacity to analyze.

The book distinguishes between three types of representational issues, depending on the type of information being represented: content versus sequence versus timing information. The representational issue for content information has been the main focus of other theorists (e.g., Anderson, 1983) and comes up here in Chapter 2: How is the basic form or content of perception and action represented? The representational issue for sequencing comes up in Chapters 3 and 4: How are the sequences of components for habitual actions and percepts represented? The third and most commonly neglected representational issue concerns timing and comes up in Chapter 5: How is rate and timing information represented in habitual actions and perceptual judgments? As we will see, the units representing sequence and timing in the node structure theory constitute control structures that are themselves hierarchically organized and activate the hierarchically organized content nodes that represent the form of perception and action. Representations of the relation between action and self-produced feedback constitute a fourth issue that other theories are forced to address but the node structure theory is not. Under the node structure theory, the other three representations for perception and action suffice to explain the representation of self-produced feedback.

Processing Themes

Four processing themes emerge at various points in the book. The main one is whether a single fundamental process (activation) adequately describes perception and action, as most other theories assume, or whether two fundamental processes (priming and activation) are required, as in the node structure theory. Related to this theme are the nature of sequential errors in speech production (Chapter 3), differences between conscious versus unconscious processes (Chapter 4), and the role of attention in modifying the weighting of different kinds of perceptual evidence (Chapters 4 and 7). As we will see, this processing theme has a direct bearing on representational issues; content nodes in the node structure theory obey a *hierarchic* principle of representation with respect to the process of activation but obey a *heterarchic* principle of representation (Kelso & Tuller, 1981) with respect to the process of priming (Chapters 3 and 4).

The second processing theme concerns the processes that enable nodes to become activated at the proper time and in the proper sequence during perception and action. Chapter 5 deals with the timing processes, and Chapters 3 and 4 with the sequencing processes.

The third processing theme concerns similarities and differences between the processes that give rise to perception versus action. The similarities are discussed in Chapters 2 and 5, which examine a set of parallelisms and interactions between perception and production. The differences are discussed in Chapter 6, which examines a set of phenomena that are fundamentally asymmetrical between perception versus action. Differences discussed in Chapter 9 between perceiving self-produced versus other-produced inputs also contribute to this theme.

The fourth processing theme concerns the frequently overlooked issue of how skill or practice influences the processes of sequencing and timing in perception and action. An example from this theme is the issue of why periodicity and systematic deviations from periodicity develop as a function of skill or practice in behaviors such as typing and handwriting (Chapter 5).

The Integration of Perception and Action

The integration of perception and action is the third major theme of the book, and in Chapter 7 this theme forms part of the larger issue of how any two heterogeneous types of information become integrated in the nervous system. Chapters 8 through 10 examine the core of the integration of perception and action, the perceptual processing of self-produced feedback. Phenomena discussed in Chapter 2 such as "rapid shadowing" and perceptual–motor adaptation also illustrate the integration of perception and action, as does the temporal incompatibility phenomenon discussed in Chapter 5, which is the curious interaction between timing mechanisms for perception and action seen in our inability to simultaneously produce and perceive rhythms with incompatible timing characteristics.

Overview of the Book

Having outlined the main themes and the intellectual context of the book, some signposts are in order regarding its chapters, their main lines of argument, and how they interconnect. In brief overview, the first half of the book develops the basics of the node structure theory, and the second half deals with applications, implications, and extensions of these basics.

Chapter 1 outlines some conceptual antecedents of the theory and defines the basic theoretical constructs that recur throughout the book: content nodes (units representing the basic components of action and perception); sequence nodes (units for determining the sequence in which content nodes become activated); and timing nodes (units for determining when and how rapidly the content nodes become activated); the basic structural properties and organizational characteristics of nodes (e.g., "more-or-less hierarchies" of content nodes within systems); their basic processing characteristics (e.g., priming and activation); and their short- and long-term memory characteristics (e.g., priming and strength of connections).

Chapter 2 focuses on the representation-of-content issue: What units represent the form or content of perception and action and how are perceptual units related to production units? I argue that some content nodes represent neither sensory experience nor patterns of muscle movement but higher level mental components common to both perception and production in speech and other cognitive skills.

The next three chapters deal with sequencing and timing in perception and action. Chapter 3 examines the sequencing of action, the question of how the components of skilled behavior become activated in proper serial order in a parallel connectionist theory, and develops detailed sequencing mechanisms for everyday skills such as producing a sentence, typing a word, or playing a phrase on the piano. Chapter 4 examines perceptual sequencing, the processes whereby perception–production units (mental nodes) perceive and register sequences of input. Chapter 4 also reviews a wide range of supporting data for a general principle of perceptual processing that follows from the node structure theory, the principle of higher level activation. Chapter 5 takes up the frequently neglected issue of timing, beginning with some general constraints on theories of timing for perception and action, and ending with a theory that meets these constraints and makes predictions for future test.

Chapter 6 points out some important asymmetries between processes in the theory that give rise to perception versus action, shows how these asymmetries explain a large number of empirical differences between perception and action, and predict some new differences for future test. Chapter 7 examines the functions of mental nodes and concludes that mental nodes evolved to enable the rapid and economic integration of many other heterogeneous sources of information in addition to perception and action. Chapter 7 also discusses problems with current explanations of visual dominance or "capture" effects and develops a new account of these effects.

The last three chapters of the book explore the processing of self-produced feedback, both in theory and in available empirical data. Chapter 8 examines self-inhibition, an inhibitory process that follows the activation of a node, and shows how self-inhibition plays a role in the processing of self-produced feedback. It also reviews supporting evidence for self-inhibition from a wide range of areas: neurophysiology, electromyography, errors in speech and typing, the misspellings of dysgraphics, the perception and recall of experimentally planted misspellings by normal individuals, and an apparently universal pattern of phoneme repetition in the structure of languages. Chapter 9 examines recent findings on the role of perceptual feedback in detecting and correcting self-produced errors, reviews the shortcomings of current theories for explaining these findings, and shows how errors are detected and corrected in the node structure theory. Chapter 10 examines how feedback in certain forms can disrupt ongoing action and discusses the constraints such disruptions provide for theories of the relationship between perception and action. A prominent example is the disruption of speech production that occurs when normal subjects hear the sound of their own voice amplified and delayed by about 0.2 s.

The Epilogue concludes the book by summarizing its main themes and concepts in a new way, analyzing the main strengths and weaknesses of the node structure theory, and pointing to fruitful directions for future research into relations between perception and action.

1 Theoretical Antecedents

> Perception cannot work by extracting production invariants if production works by using perceptual invariants. If there are perceptual invariants for use in production, then they should be used in perception.... This buck passing between production and perception has only been possible because few theorists have attempted to solve the problems of both areas within the same general theory.
>
> (Howell & Harvey, 1983, p. 203)

This chapter describes some conceptual antecedents to the node structure theory of perception and action that I develop in the remainder of the book. Conceptual antecedents to my theory stretch back to Plato, but I mention only a sample of these antecedents here, and this chapter can be viewed as a summary of my personal sources of inspiration rather than as an authoritative historical review (see also Rumelhart, McClelland, & the PDP Research Group, 1986). I begin with two general philosophical traditions that have had longstanding and profound effects on virtually all psychological and physiological thought. I then illustrate how these traditions have influenced three current theories of the relation between perception and action in general and speech perception and production in particular. I next outline a theoretical alternative to these philosophical traditions that Lashley (1951) pointed out and that constitutes a major theme of the present book. Finally, I spell out some more recent and detailed conceptual antecedents to the theory that I go on to develop in the remainder of the book.

Philosophical Antecedents

The relation between perception and action has been debated since the time of Descartes, and two general philosophical views have prevailed in this continuing debate. One is that action is subordinate to and less important than perception, and the other is that perception and action constitute separate domains of inquiry.

The Subordination-of-Action Tradition

Many philosophers have viewed action as functionally, temporally, and evaluatively subordinate to perception; functionally subordinate because they considered perception the sole means by which knowledge is acquired (empiricism), temporally subordinate because they considered perception a necessary precursor to action (paleobehaviorism), and evaluatively subordinate because they viewed the contemplative life as superior to a life of action (see Plato).

Effects of Philosophical Subordination

Effects of the conceptual subordination of action seem predictable in retrospect. The topic of perception has attracted a great deal of attention, whereas the topic of action has been relatively neglected (e.g., Gentner, 1985). Of course, psychologists often give another reason for choosing to study perception rather than action, namely that perception is methodologically easier to study. For example, Fodor, Bever, and Garrett (1974) note that psycholinguistic research has, until very recently, concentrated almost exclusively on perception, rather than on production, or on the relation between perception and production, and they attribute the neglect of production to methodological difficulties.

However, the methodological difficulty hypothesis is clearly incomplete or inadequate as a general explanation of the relative disinterest in action. Recent studies of action (e.g., Sternberg, Monsell, Knoll, & Wright, 1978) are as well controlled as any in perception. Physiology provides another problem for the methodological difficulty hypothesis. Physiologists can trigger actions electrophysiologically and thereby overcome the hypothesized methodological problems, but they too have studied perception more often than production. The methodological difficulty hypothesis also has difficulty explaining reversals in the general trend. For example, whereas perception has received more attention than production in the field at large, the opposite is true in the case of speech errors (Chapter 6). Naturally occurring misperceptions, or slips of the ear, have been collected and studied much less often than naturally occurring misproductions, or slips of the tongue (Fromkin, 1980). Finally, the whole idea that methodological ease represents a viable reason for examining or not examining a general topic area such as perception versus action seems open to question.

An Assessment of Philosophical Subordination

Contrary to the long-held philosophical subordination position, everyday perception and action interact with and support one another, and neither can be considered functionally or temporally subordinate to the other. The main function of perceptual and cognitive systems is to guide purposeful actions and to adjust ongoing actions to the situation at hand. As Allport (in press, p. 2) points out, "Perceptual systems have evolved in all species of animals solely as a means of guiding and controlling action, either present or future." Perceptual systems aren't primarily designed to describe and classify the environment in answer to a question such as "What is out there?" but to address the more general questions, "What does it signify for me? What must I do about what's out there?" (after D. M. MacKay, 1984). In short, the nature of the information required for the guidance of actions ultimately determines how perceptual and cognitive systems structure the sensory and intellectual environment. Functionally, perception is as subordinate to action as action is to perception.

Temporally also, perception sometimes precedes action, and action sometimes precedes perception. When pricked with a pin, for example, we perceive the pain only after withdrawing the finger; the awareness of pain follows rather than precedes the behavior (James, 1890). And sometimes action proceeds in the absence of perceptual awareness. For example, we continually and automatically use visual cues to orient ourselves in space, stand erect, and perform actions such as walking, even though we never perceive or become aware of these visual cues (D. N. Lee & Lishman, 1974). Usually, however, action and perception take place at the same time. When making saccades in the visual field, for example, perception and action are so intimately intertwined that temporal priority or subordination is impossible to assign.

Even from an evolutionary perspective, perception cannot take precedence over action; systems for perception and action are in general so intimately interrelated as to require mutual adaptation. Consider speech perception–production for example. The capacities for perceiving and producing speech could only have evolved simultaneously. If a mutation suddenly enabled a group of humans to understand language, their chances of surviving to transmit the mutation would only improve if a second group of humans had a mutation that enabled them to speak (see Geschwind, 1983). Moreover, the mutation that enabled this second group of humans to speak would only improve *their* chances of survival if they had a language to speak and someone to understand them when they spoke. Speech perception could not have evolved before speech production, and vice versa.

The Segregation of Perception and Action

The second major tradition in the history of philosophical ideas relating perception and action is that perception and action constitute separate systems and domains of inquiry. Beginning with Descartes, the afferent processes that mediate perception of the external world (the mind) have been considered separate from the efferent processes that mediate action in the external world (the will). These supposedly separate systems have also been assigned different functions. Perceptual systems are supposed to register and construct a meaning for sensory events, whereas motor systems are supposed to write and execute motor commands. As Turvey (1977) points out, perception and action have virtually no contact with one another in this traditional dichotomy; how a perceptual system perceives neither influences nor is influenced by how the motor system uses perception.

Effects of Philosophical Segregation

Philosophical segregation of action and perception has also had predictable effects. Two separate research areas, with little or no interaction between them, have developed in parallel, one set specializing in afferent processes, the other in efferent processes. Not just in psychology, but in the many other disciplines interested in perception and action (see the Preface), theories of action have been constructed without reference to perception, and theories of perception have been constructed without reference to action. With some notable exceptions such as the perceptual learning theory of Held and Hein (1963), virtually no theories have attempted to solve the problems of both perception and action at the same time (Howell & Harvey, 1983), and even Held and Hein's (1963) theory assumes completely separate units for perception versus action.

Mirroring the segregation tradition in philosophy, psychologists have not only chosen to study perception more often than action, they by and large have attempted to study perception in the absence of action. Perceptual experiments are characteristically designed to exclude the possibility of action, and this perception-without-action approach has almost certainly influenced the nature of the perceptual systems examined. For example, perceptual experiments have focused mainly on vision and typically attempt to eliminate action by using presentation times that are so brief as to prevent the possibility of eye movements. Touch, on the other hand, has received relatively little attention, perhaps because touch confounds the traditional dichotomy between perception and action: Movement of the hand and tactile perception of an object are cotemporaneous and cannot be separately examined or factored out in everyday tactile perception. Even though touch is phylogenetically older and more basic than either vision or speech (von Bekesy, 1967), the perception-without-action approach must avoid touch as impossible to study *in vacuo*.

Assessment of Philosophical Segregation

Needless to say, the perception-without-action approach is by definition unsuited for studying relations between perception and action. However, recent developments in many disciplines have contradicted the long-held view that perception and action are completely separate and call for a new approach to the whole topic. For example, recent neurophysiological research has made it increasingly obvious that the traditional distinction between afferent versus efferent processes in the cortex can no longer be usefully maintained. For example, Ojemann (1983) discovered cortical sites where electrical stimulation interferes with both the perception and production of everyday actions. The sensory and motor areas are inseparable in these and other studies, as if some of the units responsible for afference and efference in the cortex are identical.

Theories Incorporating the Segregation Assumption

To illustrate how the philosophical segregation of perception and action has influenced psychological theories, I examine three well-known theories: the classical theory, the motor theory of speech perception, and feedback control theory in its application to speech perception–production. Like other theories of perception–production, all three theories explicitly attempt to relate perception and action via separate rather than shared components for perception versus production. Beyond this, the three theories are remarkably different. They postulate different mechanisms, and they deal with different perception–production issues. Their only other shared characteristic is a state of crisis; all three have encountered fundamental phenomena that contradict their basic assumptions. However, I do not attempt to systematically describe or present criticisms of these theories. Nor do I compare these theories with my own. Only after I have developed the relevant aspects of my own theory do I compare it with other more recently published theories of perception (e.g., McClelland & Elman, 1986) and of action (e.g., Norman & Rummelhart, 1983).

The Classical Theory

The classical theory of the relation between speech perception and production holds that the two systems employ separate components at every level of processing (see Straight, 1980). Broca and Wernicke pioneered this theory (D. G. MacKay et al., 1987): They argued from studies of left hemisphere brain injuries that production is localized in one area of the brain and perception in another, interconnected but separate area. However, recent studies using a variety of new and more sophisticated techniques suggest that the picture is more complicated. Expressive and receptive deficits are usually commensurate in extent. For example, with appropriate controls for lesion size, aphasics with severely impaired production also display severely impaired comprehension and vice versa (Mateer, 1983). Moreover, deficits in perception and production are usually similar in nature. Production deficits tend to be more obvious than perceptual deficits in everyday life, because aphasics can simulate comprehension using nonlinguistic cues. However, when given sophisticated tests of comprehension with appropriate controls for semantic and pragmatic cues, Broca's aphasics display comprehension deficits that parallel their more readily observed production deficits. As W. E. Cooper and Zurif (1983, p. 228) point out, "Recent studies are in agreement in concluding that, to the extent that Broca's aphasics show *relatively* intact comprehension, it is largely based on their ability to utilize semantic and pragmatic cues independent of sentence structure."

The comprehension deficits of Wernicke's aphasics are likewise matched by production deficits, usually involving sentential rather than phonological units (Blumstein, 1973). Difficulties with word order are a typical problem, word salads representing the most extreme case. Although the sentential intonation of Wernicke's aphasics often sounds normal, their speech typically lacks content, contains neologistic or nonsense elements, and shows errors in sound and meaning. Contrary to the classical theory, Wernicke's aphasics are agrammatic in *production* as well as comprehension (W. E. Cooper & Zurif, 1983).

Ojemann's (1983) recent findings using cortical stimulation techniques present another problem for the classical theory. Ojemann (1983) discovered sites where electrical stimulation interferes simultaneously with the mimicking of orofacial gestures and with the perceptual identification of phonemes, as if the same units played a role in both perception and production. Finally, brain scan and cerebral blood flow studies indicate that Broca's area (which under the classical theory only becomes active during production) also becomes active during comprehension (Lassen & Larsen, 1980). All of these findings are less consistent with the classical theory than with Lashley's (1951) hypothesis that common components underlie the perception and production of speech (see also Colthart & Funnell, 1987; Meyer & Gordon, 1983).

The Early Motor Theory of Speech Perception

The early motor theory of speech perception recognized the importance of interaction between the systems for perceiving and producing speech. Motor units that are (necessarily) distinct from the perceptual units can come to the aid of the perceptual units under the motor theory. That is, speech perception and production employ separate components, but at least some speech sounds are perceived with the help of the components that are used for producing them (Liberman, Cooper, Harris, & MacNeilage, 1962; Studdert-Kennedy, Liberman, Harris, & Cooper, 1970).

As Howell and Harvey (1983, p. 215) point out, "Motor theory attempted to explain something about which very little was known (i.e., speech perception) in terms of something else about which even less was known (i.e., speech production). The problems associated with it are legion." One of the currently unresolved problems concerns the logical basis of the theory. In order for a pattern of acoustic energy to call up its appropriate production components, a full-fledged perceptual analysis is necessary (e.g., Pick & Saltzman, 1978). This brings the basis for the theory into question because a full-fledged perceptual analysis prior to motor consultation means that perceptual components can accomplish speech recognition without help from the motor components.

Feedback Control Theory

Feedback control theory is in some sense a converse of the motor theory. The perceptual mechanisms come to the aid of the production mechanisms under feedback control theory because perceptual feedback plays a direct and necessary role in producing ongoing speech or action (Adams, 1976; Schmidt, 1982). For example, under feedback control theory, auditory feedback from the *pr* in the word *production* could function to trigger production of the *o* and so on for the remainder of the word.

As expected under this and other theories, an intact auditory system is necessary for the *acquisition* of normal speech production. However, once speech has been acquired, eliminating or distorting auditory feedback has little effect on the ability to produce intelligible speech (Siegel & Pick, 1974), which suggests that sensory feedback may be unnecessary for well-practiced speech production.

Articulatory disruption does occur when auditory feedback is amplified and delayed (B. S. Lee, 1950), but even here, feedback control theory fails to fit the detailed nature of this phenomenon (reviewed in Chapter 10).

Lashley's Alternative: Common Components

Not all theoretical thinking in psychology has adopted the assumption that components for perception and action are completely separate or unshared. In particular, Lashley (1951, p. 186) proposed that speech comprehension and production make use of common components and mechanisms because "the processes of comprehension and production of speech have too much in common to depend on wholly different mechanisms."

Like Lashley, I am especially concerned in the present volume with the concept of shared components for perceiving and producing speech. After reviewing the available evidence in Chapter 2, I conclude that perception and production share some components but not others. One system of unshared components represents patterns of sensory input for perception, and another system represents patterns of muscle movement for production, while the systems of shared perception–production components represent phonological units such as segments and syllables and sentential units such as words and phrases.

However, the concept of integrated or shared perception–production mechanisms is not restricted to speech. For example, Darian-Smith, Sugitani, and Heywood (1982) discovered cells in the somatosensory cortex that respond both to finger movement and to sensory properties of a textured surface. Indeed, Pribram (1971) viewed the so-called motor cortex as a system for somatosensory regulation as well as for action, arguing that the motor representation of an action must contain a perceptual "image of achievement," because processing of feedback is necessary to ensure that the action has been executed as intended. Supporting this view, Pribram (1971) and Kornhuber (1974) reviewed evidence indicating that cells in the motor cortex are responsive to cutaneous and somatosensory stimuli from the body part moved.

Self-perceptions of actions induced by cortical stimulation are consistent with these observations. For example, when motor cortex stimulation results in arm movement, subjects never report an introspective sequence beginning with an urge to move the arm, followed by arm movement, and ending with perception of arm movement. A cortically induced movement seems unwilled, happens by itself, and is perceived and performed simultaneously (Bridgeman, 1986).

Puzzling Asymmetries Between Perception and Action

Asymmetries between perception versus production processes represent the main theoretical puzzle or challenge facing the idea of shared perception–production units and may explain in part why virtually no psychological theories have taken up Lashley's (1951) suggestion. An example of asymmetry is the fact (discussed in greater detail in Chapter 6) that speech perception can proceed much more quickly than speech production. Computer-compressed speech remains perceptually intelligible at five to seven times the rate that people can produce speech of comparable intelligibility (Foulke & Sticht, 1969). This rate asymmetry cannot be completely explained in terms of the muscular or biomechanical factors involved in speech production but reflects a central and inherent processing difference that must be explained in theories of speech perception–production (Chapter 6).

What processing differences could enable perception to proceed so much faster than production? The node structure theory postulates several fundamental processing differences (summarized in Chapter 6) that together explain not only the rate asymmetry but many other asymmetries between perception and action, such as perceptual differences between self-produced versus other-produced feedback, differential effects of practice on perception versus production, and asymmetries in the nature of the errors that occur in perception versus production.

Antecedents to the Processing Characteristics of the Node Structure Theory

The basic components of the node structure theory are *nodes*, a psychological term that dates back at least to Collins and Quillian (1969). However, my use of the term *nodes* resembles Wickelgren's (1979) and McClelland and Rumelhart's (1981) rather than Collins and Quillian's (1969), Estes' (1972), or Anderson and Bower's (1973). Nodes in the latter writings refer to intersections in a parsing tree and represent descriptive rather than theoretical terms, whereas nodes in the node structure theory are theoretical constructs, processing units that share the same structural characteristics and dynamic or processing capabilities and respond in the same way to basic variables such as practice (repeated activation). Here I discuss the dynamic characteristics of nodes, which go well beyond the concept of an intersection and have other historical antecedents of their own. The remainder of the book then examines how these dynamic characteristics become implemented during everyday perception and action.

Dynamic Characteristics of Nodes

Nodes have five dynamic properties that are relevant to all aspects of the organization of perception and action: activation, priming, satiation, self-inhibition, and linkage strength. Each of these dynamic properties, taken by itself as in the discussion below, is remarkably simple, but interactions between dynamic properties can be quite complex. Each property influences the others in complex ways that depend on the current state of the node and on its history of activity over the course of a lifetime. Also, perception and action use these dynamic processes differently in the theory, and these processing differences contribute to already observed perception–production differences, such as the maximal rate asymmetry, and predict new asymmetries for future test. Illustrating the systematicities in how the dynamic properties of nodes interact will occupy much of the remainder of the book.

Node Activation

I use the term *activation* as short for *node activation*, a process necessarily for both perception and action in the theory. My use of the term activation mirrors that of Lashley (1951) but differs from many other current uses of the same term in the cognitive and brain sciences. I attempt to avoid terminological confusion at the outset by comparing my usage with these other concepts of activation.

Like neural activation, node activation is all-or-none and self-sustained. Activation lasts for a specifiable period of time, independent of whether the sources that led originally to activation continue to provide input. However, node activation can, and in the case of mental or perception–production nodes, invariably does, involve activation of more than one neuron. Neurons and nodes also differ greatly in how long they remain activated and in their recovery time following activation. For example, neurons require at most a few milliseconds to recover from activation, whereas nodes require anywhere from a few milliseconds to hundreds of milliseconds (Chapter 8).

Node activation also differs from the concept of spreading activation in propositional network theories such as Anderson (1983). Node activation never "spreads," and its intensity never changes. Unlike spreading activation, node activation remains constant with "distance," fatigue, and the number of other nodes that an activated node is connected to. Moreover, node activation is sequential and nonautomatic in nature. A special activating mechanism must become engaged to determine when and in what order different nodes in my theory become activated. By way of illustration, the numbers in Figure 1.1 represent the typical order in which the adjacent nodes become activated during production.

During its period of self-sustained activation, a node simultaneously primes all nodes connected to it, and as we will see, priming is necessary in order to activate a node. Another characteristic distinguishing node activation from spreading activation is the occurrence of self-inhibition (discussed in a following section), a brief period of reduced excitability that follows node activation.

Node Priming

My use of the term *priming* also dates back to Lashley (1951). Priming refers to a transmission across a connection that increases subthreshold activity and prepares the connected node for possible activation. Because nodes must become primed in order to become activated, priming is a necessary precursor to all perception and action. An activated node simultaneously primes all nodes connected directly to it, and nodes that are "once removed" from an activated node also receive priming but to a lesser extent. Thus, priming falls off sharply in degree with distance from the source. An activated node primes its connected nodes

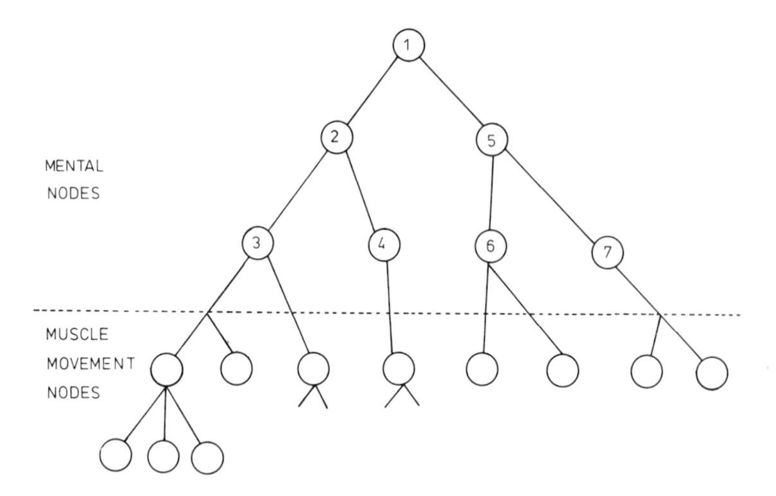

FIGURE 1.1. A top-down hierarchy of nodes for organizing an arbitrary sequence of behavior. The numbers represent the typical order in which the mental nodes become activated during production. (From "The problem of Rehearsal or Mental Practice" by D. G. MacKay, 1981, *Journal of Motor Behavior, 13*(4), p. 281. Copyright 1981 by Heldref Publishing Co. Adapted by permission.)

most strongly (first-order priming), while a node receiving first-order priming primes its connected nodes less strongly (second-order priming). Third-order priming arising from the activation of a single node is negligible in degree, and unless it summates with priming from other sources, third-order priming can be ignored in theories of production. Thus, priming spreads, but only to a limited degree, and unlike propositional network theories, activation never spreads at all.

Priming summates across all simultaneously active connections (spatial summation), and priming accumulates during the time that any given connection remains active (temporal summation). Consider, for example, the temporal summation of top-down priming across the connections illustrated in Figure 1.1. Node 1 becomes activated first and simultaneously primes nodes 2 and 5 (Figure 1.1). However, node 5 cannot become activated until nodes 2, 3, and 4 have been activated. Priming of node 5 therefore continues to summate during the time that nodes 2, 3, and 4 are being activated. The anticipatory nature of this accumulating priming facilitates the eventual activation of node 5 and all other "right-branching" nodes at every level in such a hierarchy. Because virtually all top-down connections are divergent (one-to-many), top-down anticipatory effects are universal in the theory (D. G. MacKay, 1982). As we will see, however, these anticipatory effects incur a built-in cost. Temporal summation of priming increases the probability of "anticipatory errors," where an about-to-be-produced component is produced before its time, the most common class of errors at either the phonological or sentential levels of speech production (e.g., Cohen, 1967).

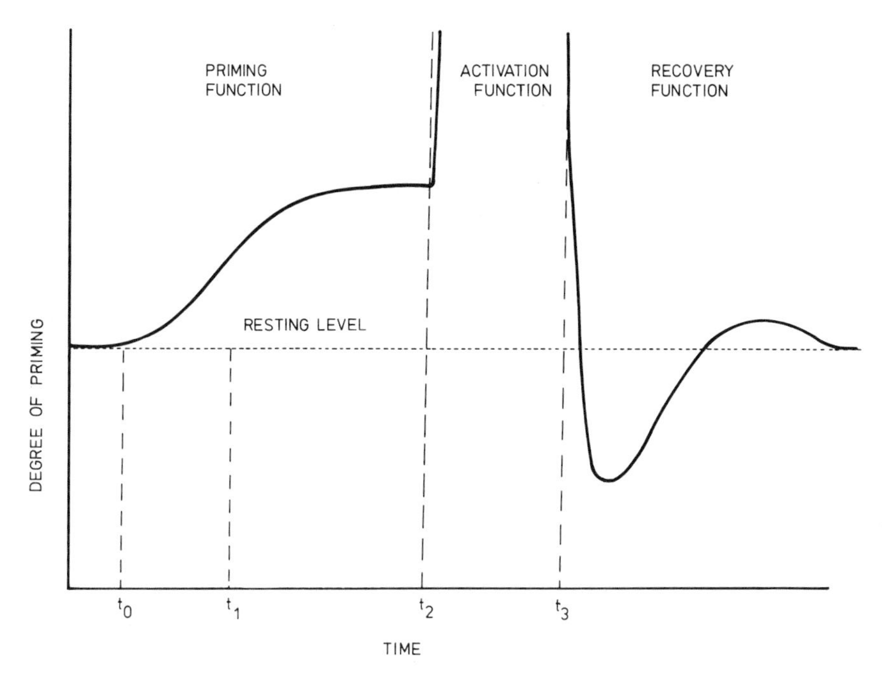

FIGURE 1.2. The priming, activation, and recovery phases for a single node. The priming function shows how priming summates to asymptote following onset of priming at t_0. The activation function illustrates multiplication of priming and self-sustained activation until time t_3. The recovery cycle shows how priming first falls below resting level (self-inhibition) and then rebounds (the hyperexcitability phase) following termination of activation at time t_2.

Priming only summates to some subthreshold asymptotic level (Figure 1.2), and cannot by itself cause a connected node to become activated. As a consequence, priming never results in behavior when the lowest level muscle movement nodes representing an action become primed.

Unlike activation, priming cannot be self-sustained and begins to decay as soon as the activity of a connected node stops (Figure 1.2). Also unlike activation, priming is not followed by a period of self-inhibition and recovery. Finally, priming is untimed and order free. No special triggering mechanism is required to determine when and in what order nodes become primed. In this sense, priming is automatic and parallel in nature, whereas activation is controlled and sequential.

Despite these many differences, activation and priming are intimately related in the theory. A minimal degree of priming is necessary for activation. Unless a node achieves this minimum priming level, designated the *commitment threshold*, its activation mechanism will be unable to activate it. However, achieving the minimal level of priming is insufficient to guarantee activation if and when the activation mechanism is applied. In order to become activated, a node must also receive more priming than any other node in its domain. Chapter 2 discusses the theoretical and empirical basis for this "most-primed-wins" principle, the means by which all nodes become activated.

Self-Inhibition

After nodes representing the components of skilled behavior become activated, they undergo a brief period of self-inhibition, during which their level of priming falls below normal or resting level (Figure 1.2). Following self-inhibition, nodes undergo a recovery cycle consisting of a hyperexcitability phase followed by a return to resting level. During the hyperexcitability phase, or postinhibitory rebound, priming first rises above and then returns to resting level (Figure 1.2). Various sources of evidence bearing on the time characteristics of self-inhibition and the recovery cycle are discussed in Chapter 8.

Satiation

Satiation refers to a fatigue process during which a node becomes less responsive to priming. This reduced sensitivity occurs when the node has been activated repeatedly over a prolonged period of time, say 5 to 30 s, and manifests itself in two ways. The duration of self-inhibition following activation becomes extended, and the rebound from self-inhibition falls below normal or resting level. Satiation of course varies in degree, depending on the extent and duration of repeated activation.

Linkage Strength

The concept of linkage strength also has a long and distinguished history, dating back at least to Thorndike (1898). Linkage strength represents a relatively long-term characteristic of a connection that has been used to explain a wide range of practice effects in the psychological literature (D. G. MacKay, 1982). Practice, or more specifically, the frequency with which a node has been primed and activated via a particular connection in the past determines linkage strength in the node structure theory. However, linkage strength has a special relation to priming in the theory. Connections with high linkage strength transmit priming more rapidly and provide more priming at asymptote than do connections with low linkage strength. That is, linkage strength influences how fast priming will summate across a connection per unit time (represented by the initial slope of a priming function, such as the one illustrated in Figure 1.2), and linkage strength influences how much priming can be transmitted across the connection before asymptote is reached.

In summary, the dynamic properties of nodes (activation, self-inhibition, priming, satiation, and linkage strength) are closely interrelated. Priming is necessary for activating a node, and the degree of priming is related to the probability of activation in ways discussed in subsequent chapters. Activating a node increases the linkage strength of its connections and causes its connected nodes to become primed. Linkage strength in turn influences how much and how rapidly priming can be transmitted across a connection. Finally, self-inhibition terminates activation and, during satiation, is itself influenced by activation.

Special Dynamic Properties

In addition to the universal dynamic properties discussed previously, some nodes also have special dynamic properties, such as quenching, multiplication of priming, and periodicity. These special dynamic properties differentiate three classes of nodes: content nodes, sequence nodes, and timing nodes. Content, sequence, and timing nodes also differ in how they connect with other nodes and in the functions they perform in perception and action. *Content nodes* represent the form or content components of an action or perception. *Sequence nodes* activate content nodes in some specifiable sequence. And *timing nodes* activate sequence nodes at some specifiable rate.

Quenching

Quenching is a special characteristic of content nodes, with conceptual antecedents in Grossberg (1982). Once a content node becomes activated, it quenches or inhibits the sequence node that originally caused it to become activated. The next chapter discusses reasons for including this quenching mechanism in the theory.

Multiplication of Priming

Multiplication of priming is another process anticipated in some respects by Grossberg (1982) and represents a special function carried out by sequence nodes. An activated sequence node doesn't simply prime its connected nodes; it multiplies their existing level of priming by some factor per unit time. For example, a sequence node might double the level of priming of a connected content node every 2 ms. This multiplication of priming process provides the basis for the most-primed-wins principle by which all content nodes become activated.

Periodicity

Periodicity is a process with a long theoretical history and refers in the node structure theory to an endogenous and inherently rhythmic pattern of activation that is characteristic of timing nodes. Once a timing node becomes engaged, it spontaneously self-activates every (say) 10 ms.

2
The Structure of Perception and Action

> I have devoted so much time to the discussion of the problem of syntax, not only because language is one of the most important products of human cerebral action, but also because the problems raised by the organization of language seem to me to be characteristic of almost all other cerebral activity. There is a series of hierarchies of organization; the order of vocal movements in pronouncing the words, the order of words in the sentence, the order of sentences in the paragraph, the rational order of paragraphs in a discourse.
>
> (Lashley, 1951, pp. 121–122)

In general approach, the present book reverses the traditional strategy discussed in Chapter 1 of treating perception and action separately, because I begin with the evidence for shared perception–production units, which play a role in both perception and action. By taking up action next, the book also reverses the traditional strategy of giving priority to perception. I attempt first to specify a detailed set of theoretical processes for sequencing and timing the production of speech and other skilled behaviors involving shared perception–production components. I then examine how these shared perception–production components give rise to perception, and I develop a theory with applications to classical perceptual problems such as categorical perception, perceptual invariance, the nature of perceptual errors, perception of the distal stimulus, perception of sequential inputs, and the problem of ambiguity in perception.

My ultimate goal is a unified and general theory, unified in the sense of dealing with all aspects of dynamic or on-line perception–production, and general in the sense of dealing with these dynamic aspects at all levels, including, in the case of speech, the muscle movement, phonological, and setential levels. Along the way, I review a wide range of empirical findings from various domains of inquiry (mainly cognitive psychology, neuropsychology, psycholinguistics, cybernetics, and motor control), but my main aim throughout is to develop the new theory in as detailed a manner as possible.

To facilitate exposition, I develop the theory in stages corresponding to issues raised in Chapter 1. What are the common components that perception and production systems share? How do these common components function in a theory of sequencing and timing in speech production? What processes involving these common components give rise to perception? How can asymmetries between perception and action be explained in a theory incorporating shared perception–production components? What functions did common perception–production components evolve to serve? And what role does perceptual feedback play in ongoing action? The present chapter addresses the first of these issues, Lashley's (1951) hypothesis concerning shared units for perceiving and producing speech.

The Mental Node Hypothesis

The mental node hypothesis is the cornerstone of the node structure account of the relationship between perception and action. Under the mental node hypothesis, some of the nodes for perception and production are identical. These mental nodes or shared perception–production units represent neither sensory experience nor patterns of muscle movement but higher level cognitive components common to both perception and production (see also the "hidden units" of Rumelhart, McClelland, & the PDP Research Group, 1986). By definition, mental nodes are neither purely motor nor purely sensory but both, and they become active during perception, production, and cognition (e.g., internal speech). For example, mental nodes in the language modality represent phonological units, such as segments and syllables, and sentential units, such as words and phrases.

However, not all of the components for speech perception–production are shared. The basilar membrane and associated auditory pathways register speech inputs but play no role in speech production, for example. Nor do the muscles for the respiratory, laryngeal, velar, and articulatory organs contribute to speech perception. Here, then, are two separate systems that do not share both perceptual and production functions. One system contains sensory analysis nodes, which represent the patterns of auditory input. The other system contains muscle movement nodes, which represent the patterns of muscle movement for producing speech sounds.

The hypothesis at issue is whether a common set of nodes becomes primed when we perceive a word (or sentence) and when we produce it, either aloud, or within the imagination (internal speech). Although I focus on examples from speech here, this mental node hypothesis is intended to apply not just to speech, but to all systems for everyday action and perception. A common set of mental nodes is assumed to be involved, for example, when a chess player perceives and comprehends a sequence of chess moves or generates the same sequence of moves either on the board or within the imagination. The mental nodes for comprehending and generating chess moves are of course distinct from the sensory nodes that analyze the visual pattern of the chess board and from the motor nodes that generate the sequence of muscle contractions for moving the pieces.

For readers interested in other (nonspeech) perception–action systems, D. G. MacKay (1985) discusses the mental nodes involved in hammering a nail, shifting gears in a standard gear-shift automobile, and the generation of Morse code.

Figure 2.1 provides a general overview of the mental node hypothesis. The mental nodes send "top-down" outputs to the muscle movement nodes during production and receive "bottom-up" inputs from the sensory analysis nodes during perception. These sensory analysis nodes also analyze self-generated perceptual feedback, represented by the broken line in Figure 2.1. In what follows I first specify the types and structure of connections between mental nodes and review various sources of evidence for mental nodes. I then explore some implications of the mental node hypothesis for the nature of interactions between the perception and production of speech. Finally, I conclude the chapter with some limitations and possible extensions of the mental node hypothesis.

Types of Mental Nodes

Mental nodes fall into three functional classes based on their dynamic properties (discussed in Chapter 1) and on the structure of their connections with other nodes. *Content nodes* represent the form or content components of an action or perception; *sequence nodes* represent the order in which content nodes become activated; and *timing nodes* determine when to activate the sequence nodes, which in turn activate the content nodes. All three types of nodes normally play a role in both perception and production. However, I focus here on the structure

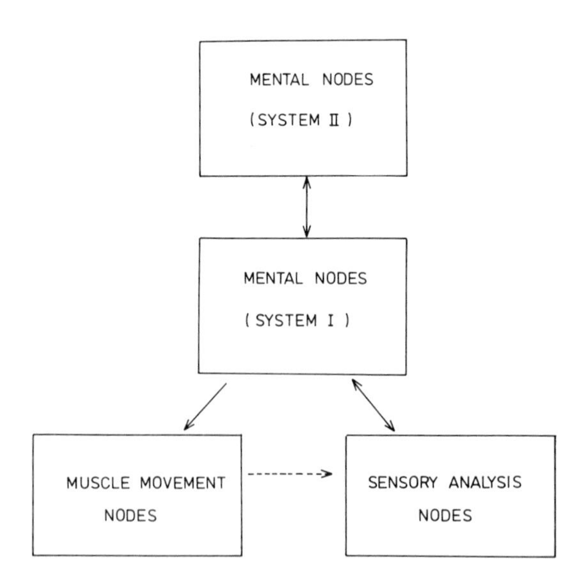

FIGURE 2.1. An overview of the mental node hypothesis. The solid arrows represent internal connections between mental nodes, muscle movement nodes, and sensory analysis nodes, while the broken arrow represents self-generated feedback.

of connections between content nodes. Indeed, when I use the term *node* in the remainder of this chapter, I refer to content nodes. I discuss sequence and timing nodes and how they interconnect with content nodes, in the subsequent chapters on processing.

The Structure of Connections Between Mental Nodes

The top-down connections between mental nodes can be described as "more-or-less hierarchic," rather than "strictly hierarchic." To illustrate this distinction, I will begin by analyzing a strict hierarchy and then discuss why, in general, top-down connections only form more-or-less hierarchies.

Top-down connections between the nodes representing the sentence "Theoretical predictions guide research" (Figure 2.2), provide an example of a strict

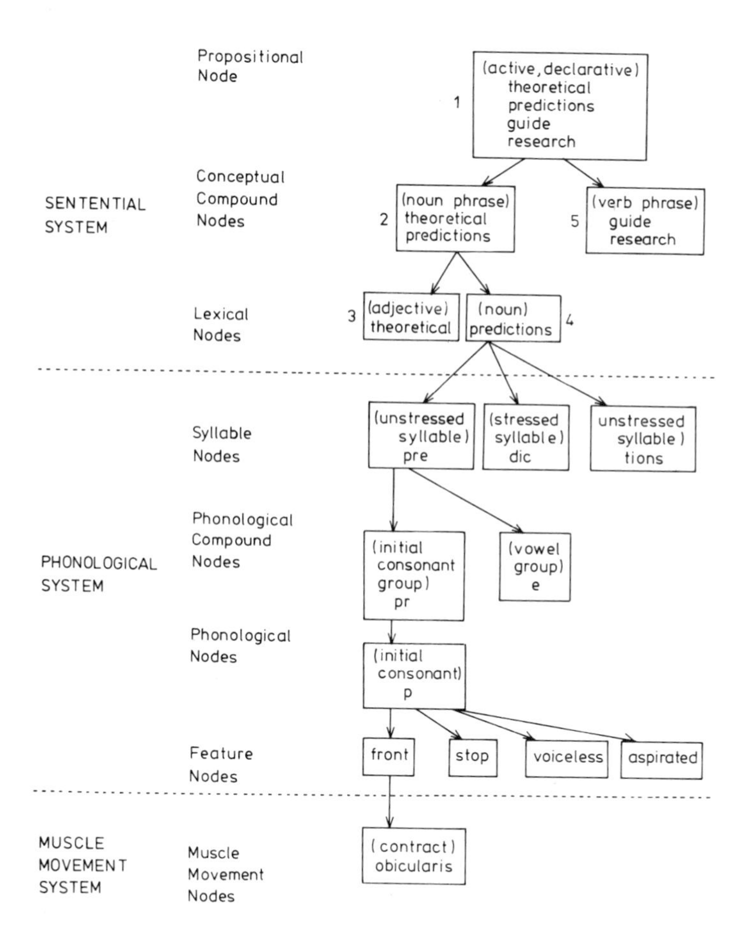

FIGURE 2.2. A sample of top-down connections for producing the preplanned sentence "Theoretical predictions guide research."

hierarchy. Following a notational convention developed in D. G. MacKay (1982), I refer to particular nodes by means of a two-component label: the content that the node represents appears in italics, followed immediately in parentheses by its sequential domain (explained later). The significance of this two-component label will become apparent when I discuss activating mechanisms in Chapter 3. Thus, the highest level node representing the entire thought underlying the sentence in Figure 2.2 has the content "Theoretical predictions guide research," occurs in the domain (active declarative), and is labeled *theoretical predictions guide research*(active declarative). This particular node is connected to two other nodes, labeled *theoretical predictions*(noun phrase) and *guide research*(verb phrase) (Figure 2.2). *Theoretical predictions*(noun phrase) is connected with two lexical nodes, *theoretical*(adjective) and *predictions*(noun). These lexical nodes are connected with specific phonological nodes, representing syllables (e.g., *pre*); phonological compounds (e.g., *pr*); segments (e.g., *p*); and features (e.g., the one representing the frontal place of articulation of *p*). Later in the chapter, I discuss some of the data supporting the particular units and connections illustrated in Figure 2.2, but the reader is referred to D. G. MacKay (1972; 1973b; 1978) and Treiman (1983) for details of the full range of supporting evidence. Numerals next to each node illustrate order of activation during production.

A more complex but otherwise similar hierarchy of nodes is assumed to underlie the control of muscle movements, but so little is known about the detailed nature and structure of connections within the muscle movement system for speech, or any other action system, that such a hierarchy cannot be represented here. Figure 2.2 illustrates nothing of this hierarchy of connections and indicates only one of the hundreds of muscle movement nodes that must become activated in producing the sentence "Theoretical predictions guide research." I simply do not know what all of the remaining muscle movement nodes are, let alone the structure of their interconnections; and even if I did, this information would be too complex to include in a form resembling Figure 2.2.

Hanging Branches and More-or-Less Hierarchies

Top-down hierarchies are in general only more-or-less hierarchies because some top-down connections in some node structures do not go all the way to the ground (the lowest level muscle movement nodes that give rise to behavior). These "hanging branches" are connections that exist but are not used for generating behavior in the current context. Because hanging branches do not cause their connected nodes to become activated, they represent a break in the hierarchic chain of command leading to behavior.

Hanging branches occur whenever context automatically determines the choice between two or more highly practiced response alternatives. Context-determined response specification is a very general phenomenon that occurs at all levels of a response hierarchy (D. G. MacKay, 1982; 1983), and because the mechanism is the same at all levels, I have chosen a higher level example from D. G. MacKay (1982) for purposes of illustration; the contextually determined specification of the definite versus indefinite article in English. Figure 2.3 shows the top-down connections for producing the noun phrase "the theory," in the sentence "The theory proved helpful." The node representing this noun phrase can be coded *a/the theory*(noun phrase) (see D. G. MacKay, 1982, for supporting arguments). That is, the information "definite versus indefinite determiner" isn't represented directly at the noun phrase level, but becomes specified at a lower level with the help of contextually available information. In the example under consideration, context specifies whether or not the theory in question has already been mentioned in the ongoing conversation and thereby determines the appropriate response alternative, *the*.

What is the mechanism underlying context-dependent response specification? In this particular example, the mechanism works as follows: *A/the theory*(noun phrase) is connected to *theory*(noun), and to both determiner nodes, *a*(determiner), and *the*(determiner). Each of these determiner nodes also receives a connection from another source. The other source for *a*(determiner) is a node representing the concept "new or never previously mentioned," whereas the other source for *the*(determiner) is a node representing the concept "old or

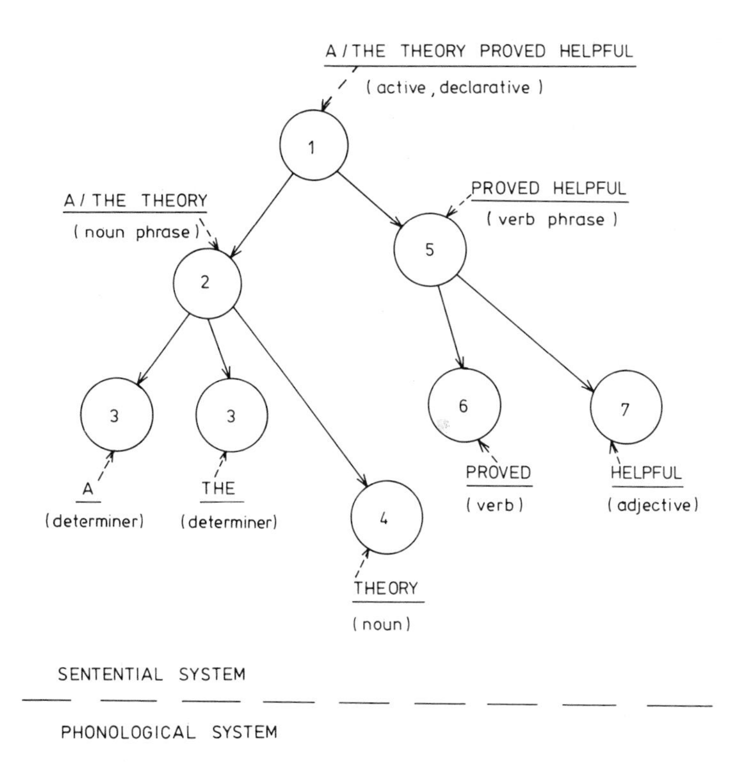

FIGURE 2.3. The top-down connections for producing the noun phrases "the theory," and "a theory," in the sentences "The theory proved helpful" and "A theory proved helpful."

previously mentioned." Because *a*(determiner) does not receive priming from its contextual source, whereas *the*(determiner) does, *a*(determiner) receives less priming than *the*(determiner), and so cannot become activated under the most-primed-wins principle when the activating mechanism is applied to the domain of determiner nodes. The connection to *a*(determiner) therefore represents a hanging branch, because it exists but is not used for generating behavior in this particular context. In general then, the most-primed-wins principle acts as an either–or gating mechanism so that when nodes in the same domain receive simultaneous priming, only the node receiving most priming from whichever (e.g., contextual) sources will become activated.

The extensiveness of context-dependent response specifications remains to be determined. For example, a similar contextual priming process could in principle select between the nouns *solely* versus *totally* in a context such as "He was solely/totally responsible for that." The process of context-dependent response specification could also help resolve the longstanding debate over the coexistence of syllable and morphological units in speech production. The debate revolves around the fact that morphemes and syllables are non-isomorphic at the surface level. Two different morphemes can map onto the same syllable, and two different syllables can map onto the same morpheme. For example, in the words *incapable* and *imprudent*, two different syllables, /in/ and /im/, represent the same negative prefix. This non-isomorphism between syllables and morphemes has led some to argue that either morphemes are a unit, or syllables, but not both. However, contextual specification via the most-primed-wins principle enables the hierarchic organization of units that are non-isomorphic at the surface level. It is perfectly possible for both syllables and morphemes to be units within the node structure theory. Both morpheme and syllable nodes represent abstract concepts, rather than surface elements per se and can connect to several different phonological nodes in the same domain. Lower level contextual sources of priming then determine which of these same-domain alternatives becomes activated. In the example under consideration, the contextual source of priming that determines whether /im-/ versus /in-/ gets produced is the place of articulation of the subsequent consonant. Moreover, many other contextually determined phonological modifications, alternative plural forms (/s/ versus /z/ versus /ez/, as in *lips*, *lids*, and *lunches*) and past tense forms (/t/ versus /d/ versus /ed/ as in *chipped*, *proved*, and *cheated*) (see Heffner, 1964, for other examples) could be determined in the same way (D. G. MacKay, 1983). For example, the lexical node for the word *proved* could connect to a syllable node, *prove*(stressed syllable), and a node representing the archiphoneme /D/, which represents the past tense abstractly by connecting with both nodes in the (voicing) domain, +*voice*(voicing), and −*voice*(voicing), as well as the other phonological feature nodes for producing /t/ versus /d/. Thus, contextual priming from the preceding consonant determines whether the −voice of /t/ or the +voice of /d/ gets activated under the most-primed-wins principle.

Action Hierarchies

An action hierarchy consists of all of the nodes that become activated in producing a preplanned behavior, including the full set of activated muscle movement nodes. Figures 2.2 and 2.3 illustrate aspects of what is, and is not, included in an action hierarchy. Figure 2.2 includes only (but not all) aspects of the action hierarchy for producing the preplanned sentence "Theoretical predictions guide research." However, Figure 2.3 includes more than just (aspects of) the action hierarchy for producing the sentence "The theory proved helpful." Nodes that receive first-order priming but do not become activated are not part of an action hierarchy, and because the hanging branch, *a*(determiner), does not become activated, it is not part of the action hierarchy for producing this particular sentence. Action hierarchies are therefore real or strict hierarchies, and not more-or-less hierarchies, and in general fail to represent the full structure of top-down connections between any given pair of nodes in the network.

Can we expect to find anatomical or neurophysiological structures in the brain that resemble action hierarchies such as the one in Figure 2.2? The likelihood of finding such structures using current technology is extremely remote. Action hierarchies are defined not by structure alone but by the occurrence of a process (activation), and we currently lack physiological definitions of either activation or priming, which would allow us to physiologically distinguish an action hierarchy from its hanging branches. Other structures must also be distinguished: the sequence nodes for activating the content nodes, the timing nodes, and other content nodes contributing connections, sometimes from other modalities. As illustrated later in the chapter, a single lexical content node typically receives connections not just from within the language modality but from many other visual, sensory, and conceptual modalities as well. And even if we could distinguish these other connections from the action hierarchy itself, anatomical action hierarchies will not be as neatly laid out as Figure 2.2. The brain lacks the systematic spatial arrangement that has been built into Figure 2.2 for ease of presentation, with the left-to-right dimension representing the order in which nodes become activated, and the up–down dimension representing the direction of priming.

Perceptual Hierarchies

Perceptual hierarchies are the input analogues of action hierarchies. They include all and only the nodes that become activated in perceiving a unitary input sequence. Nodes that only become primed, but not activated, are not part of a perceptual hierarchy. Figure 2.4 shows a typical perceptual hierarchy.

You will note that Figure 2.4 contains no sensory analysis nodes. I can say very little about how sensory analysis nodes for auditory inputs are connected to one another. Like top-down hierarchies of muscle movement nodes, bottom-up hierarchies of sensory analysis nodes are extremely complex and diverse, and the

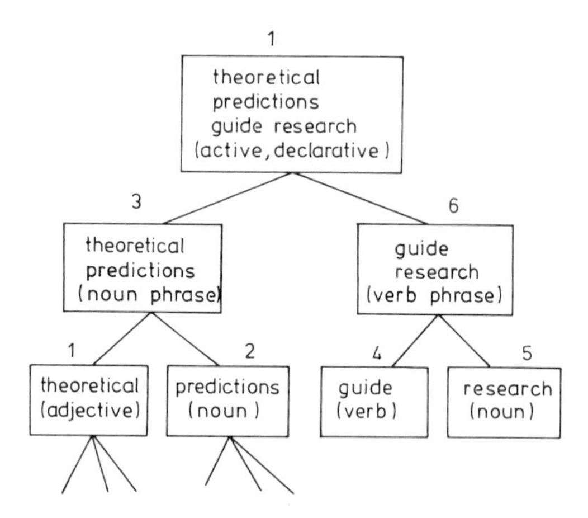

FIGURE 2.4. The perceptual hierarchy for normal comprehension of the sentence "Theoretical predictions guide research." Numerals next to each node illustrate the most likely order of activation during comprehension.

structure of their interconnections is currently unknown. For example, Lisker (1978) was able to catalogue 16 acoustic differences that could serve to distinguish a single phonological feature (the voicing of /p/ versus /b/) in a single phonological context (the words *rabid* versus *rapid*). Which sensory analysis nodes represent these acoustic differences? And what is the structure of interconnections between these nodes? All we can currently say is that acoustic analysis nodes deliver priming to phonological nodes.

This gap in our knowledge is unfortunate, but may not be especially important for an analysis of perceptual hierarchies. As we will see when I discuss perceptual processes in Chapter 4, perceptual hierarchies are quite flexible and only sometimes include sensory analysis nodes. In general, sensory analysis nodes only become primed, not activated, during everyday sentence perception. This means that sensory analysis nodes are not part of the perceptual hierarchies for normal sentence comprehension, because perceptual hierarchies only contain nodes that become activated and not just primed. Indeed, I will argue that even phonological nodes do not become activated during everyday sentence comprehension, so that Figure 2.4 represents the complete perceptual hierarchy for normal comprehension of the sentence "Theoretical predictions guide research." Numerals next to each node illustrate the most likely order of activation during comprehension.

Any given perceptual hierarchy represents only a small part of the network of bottom-up connections that become primed during perception of a unitary input sequence such as a sentence. Unlike action hierarchies, which *sometimes* represent more-or-less hierarchies, perceptual hierarchies *always* represent more-or-less rather than strict hierarchies. Hanging branches are not part of the perceptual hierarchy currently undergoing activation and constitute a universal characteristic of perception. Every activated node in a perceptual hierarchy primes many connected nodes that do not become activated. Bottom-up connections within the phonological system can be used to illustrate these hanging branches. Consider, for example, the syllable *pre* in the word *predictions*, illustrated in Figure 2.4. The syllable node, *pre*(unstressed syllable), has bottom-up connections not just with *predictions*, but with lexical nodes representing every word containing the syllable *pre*: for example, predominant, preformed, prepare, prehistoric . . . hanging branches all.

The Network in Overview

What can we say about the overall network of mental nodes for language perception–production? The existence of hanging branches means that the flow of information in the theory is neither strictly hierarchical, nor strictly heterarchical in nature. The overall network is structurally heterarchic, but functionally hierarchic. Structurally, everything can be said to connect with everything else via some relatively small number of connections in the node structure theory. The flow of priming automatically follows these existing connections and is therefore multidirectional or heterarchic in nature during both perception and action.

Functionally, however, the network is hierarchic. The activation process transforms the heterarchical connections of the overall network into local hierarchies that represent the functionally essential structures for perception and action. The next chapter discusses in detail how this activation process works: The present chapter on structure only shows that these hierarchies are there for potential use.

In the remaining chapters of the book, I argue that the heterarchic characteristics of the theory overcome the disadvantages of strictly hierarchic theories, which postulate a unidirectional flow of information but fail to explain the functional plasticity of behavior. I also argue that the hierarchic characteristics of the theory overcome the disadvantages of strictly heterarchic theories, which postulate a multidirectional flow of information but are too flexible to enable sequentially ordered action (see also Kelso & Tuller, 1981).

Systems of Nodes for Perception and Action

Functionally, nodes are organized into systems, but again only with respect to the process of activation. By definition, nodes organized into one system can be activated independently of the nodes organized into another system, and the next two chapters discuss the activation mechanisms that determine this functional organization of nodes into systems.

Figure 2.2 illustrates nodes within three different systems for perceiving and producing speech: the (speech) muscle movement system, the phonological system, and the sentential system. Activating sentential system nodes without activating nodes in the other two systems results in sequentially organized thought. Activating nodes in both the sentential and the phonological system, without activating nodes in the muscle movement system, results in internal speech (D. G. MacKay, 1981). Activating nodes in all three systems at once results in fully articulated speech. Readers interested in analogous systems of nodes for producing everyday actions such as shifting gears in a car and carrying out a preplanned shopping trip are referred to D. G. MacKay (1985).

Modalities for Perception and Action

Functionally, systems of nodes are organized into modalities, once again, via the process of activation. Nodes organized into one modality can be activated independently of nodes organized into another modality. For example, the language modality includes the language comprehension systems (including connections from the basilar membrane), and the language production systems (including connections to the lungs, larynx, velum, and articulatory organs for speaking). Modalities can also contain modalities. In the case of someone who knows two languages, for example English and Ameslan (American sign language), the language modality can be said to contain an English modality and an Ameslan modality.

Systems can participate in more than one modality, and the traditional sensory organs and pathways for vision, touch, hearing, smell, and taste all participate in several modalities. For example, the basilar membrane participates in one modality when listening to speech and in another modality when comprehending complex auditory concepts such as a police siren or a familiar musical stanza. Similarly, the retina participates in one visual modality when we comprehend a printed page and in another visual modality when we comprehend complex visual concepts such as houses and trees.

Motor end organs also participate in many different modalities, as when the tongue is used for speaking versus chewing, for example. Systems of mental nodes can likewise participate in several modalities. When producing Ameslan, for example, virtually the same sentential nodes as for English become engaged, but systems for producing hand and body movements become engaged instead of the phonological and speech muscle movement systems (D. G. MacKay, 1982).

Because different modalities interconnect extensively, nodes in one modality regularly prime connected nodes in other modalities. What makes a modality modular is that its nodes can be *activated* independently from nodes in other modalities (D. G. MacKay et al., 1987).

The McGurk effect can be used to illustrate how the modalities for speech and vision can interact via priming but are independently activated. McGurk and MacDonald (1976) had subjects listen to and observe a video recording of a person saying simple syllables, their task being to identify the syllables. The auditory syllables were dubbed in synchrony with the speaker's lip movements, but the auditory syllables sometimes differed from the lip movements. The subjects' task was to say what syllable they *heard*, and the results showed that visual features such as lip closure exerted a strong effect on what phoneme the subjects reported hearing. With a conflict between visual /pa/ and auditory /ta/, for example, subjects usually reported hearing /pa/ rather than /ta/. Apparently the visual modality nodes representing facial gestures such as lip closure are connected to and prime their corresponding phonological nodes in the language modality, and thereby influence which segment node receives most priming and becomes activated under the most-primed-wins principle. However, the visual modality nodes do not themselves become activated and give rise to perception; the subjects were unaware that visual events contributed anything whatsoever to their perception (McGurk & MacDonald, 1976).

Evidence for Mental Nodes

Many findings can be seen to support the mental node hypothesis previously discussed. Here I briefly mention four very general classes of phenomena, leaving more detailed evidence and predictions for later in the book.

Parallel Empirical Effects

As expected under the mental node hypothesis, many variables have parallel effects on perception and production. Practice is one of these variables. Repetition facilitates both production (D. G. MacKay, 1982) and perception; even recognition and discrimination thresholds for sensory qualities improve as a function of practice (Woodworth, 1938).

Complexity is another variable with parallel effects on both perception and production. By way of illustration, consider the time to perceive and produce simple (one-syllable) versus complex (two-syllable) words. On the perception side, two-syllable words are harder to identify than one-syllable words with the same frequency of occurrence, the same length in letters, and the same initial segment(s). Spoer and Smith (1973) tachistoscopically presented one- versus two-syllable words and found that subjects took longer to identify the two-syllable words (e.g., *paper*) than the one-syllable words (e.g., *paint*).

On the output side, Klapp, Anderson, and Berrian (1973) likewise presented subjects with one- versus two-syllable words controlled for initial segment(s) and length in letters, but this time the subjects' task was simply to read the words aloud as quickly as possible. The dependent variable was production onset time, the time from visual presentation of the words until acoustic onset of the subject's output. The two-syllable words required slightly (15 ms) but significantly longer onset times. To rule out a perceptual interpretation of this complexity effect, Klapp et al. had subjects produce the same words in a picture-naming task, and again, production onset time was longer for two-syllable than one-syllable words. This control finding implicates a production effect, rather than a purely perceptual effect, because number of syllables is only relevant to saying the words in this condition; the input involved pictures, which do not have syllables.

In summary, complexity has parallel effects on the input and output side, and these parallel effects are readily explained under the mental node hypothesis, where two-syllable words involve more underlying nodes than one-syllable words both in perception and in production. By way of illustration, Figure 2.5 compares the mental nodes for producing *court* and *color*, words that have identical initial segments and identical length in letters but differ in number of syllables. However, more mental nodes become involved in perceiving and producing the two-syllable word *color* than the one-syllable word *court* (see Figure 2.5). Needless to say, the parallel effects of complexity on perception and production could have arisen independently in separate rather than shared node structures, but this view requires a separate explanation for the independent emergence of these parallel structures.

The mental node hypothesis also generates some new and more refined predictions concerning the relation between production onset time and the structure of words and syllables. Two factors contribute to production onset lags under the node structure theory. One concerns the set of content nodes that must become activated before the first muscle movements for producing a word or action can begin. Because activation takes time, the more underlying nodes that must be activated, the longer will be the lag that precedes production onset. This factor

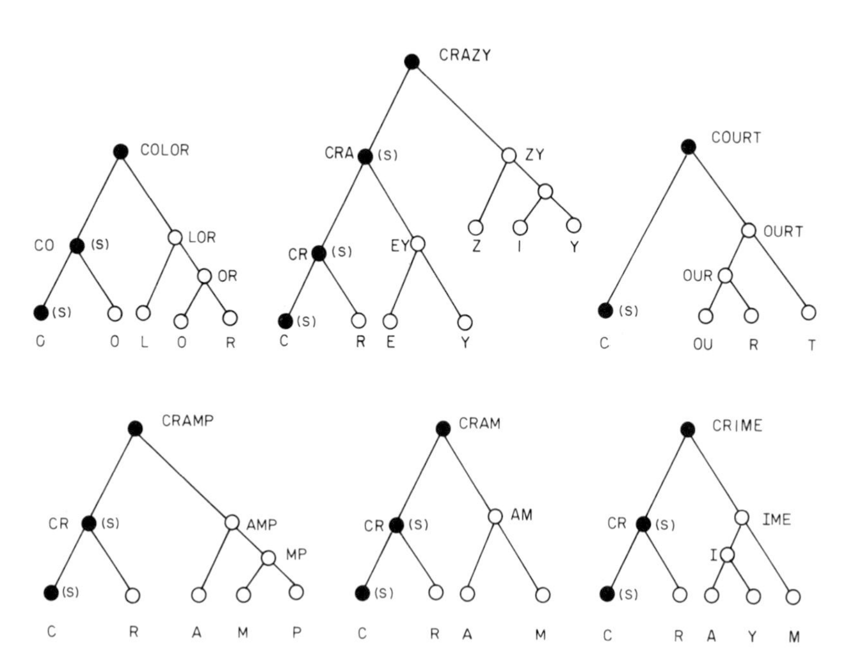

FIGURE 2.5. The structure of mental nodes for producing the words *color*, *crazy*, *court*, *cramp*, *cram*, and *crime*. Note that *crime* and *court* have equivalent length in letters, phonemes, and syllables but that *crime* and *color* have different lengths in syllables, while *cram* and *cramp* have different lengths in phonemes. (S) indicates the occurrence of a sequential decision, and filled circles indicate nodes that must be activated prior to activating the first segment node of these words.

by itself is sufficient to explain why production onset times are longer for two-syllable words than for one-syllable words that begin with the same initial segments. More mental nodes must become activated before the first segment node of a two-syllable word can become activated.

The other factor contributing to production onset lags concerns the number of sequential decisions that are required before the first muscle movements for producing the word can begin. As discussed in Chapter 3, a sequential decision is required whenever two or more nodes in different domains receive first-order priming beginning at exactly the same time. These sequential decisions take time, and the more sequential decisions that must be made, the longer the production onset time.

A detailed examination of these theoretical factors influencing production onset time leads to some new and counterintuitive predictions. The theory predicts *different* onset times for some word pairs that are equivalent in length and predicts *equivalent* onset times for other word pairs that differ in length, as measured in either syllables or segments. By way of illustration, Figure 2.5 compares the sequential decisions and mental nodes for producing the words *color, court, crime, crazy, cram,* and *cramp.* The letter *S* in parentheses (S) indicates a sequential decision, and shading indicates which nodes must be activated before the first segment node of these words can become activated. Note that *crime* and *court* have equivalent length, whether measured in letters, phonemes, or syllables. However, more sequential decisions must be made and nodes activated prior to activating the first segment node of *crime* than of *court* (Figure 2.5). The theory therefore predicts longer production onset times for words such as *crime*, which have an initial consonant cluster, than for otherwise similar words, such as *court*, which do not.

Now consider *cram* and *cramp.* These words differ in number of letters and phonemes, but they do not differ in the number of mental nodes that must be activated before the first segment node. *Cramp* only becomes more complex than *cram* after *c*(initial consonant) has become activated (Figure 2.5). In short, the theory predicts equivalent production onset times when differences between node structures arise *after* activation of the first segment node, all other factors being equal. *Crazy* and *cramp* illustrate a case where not all other factors *are* equal. As a two-syllable word, *crazy* requires more sequential decisions than *cramp* before activation of *c*(initial consonant).

Finally, consider *crime* and *color.* These words differ in number of syllables but *not* in sequential decisions and number of mental nodes prior to activation of the first segment node (Figure 2.5). The theory predicts *identical* production onset times when equivalent node structures and sequential decisions precede activation of the first segment node in one- versus two-syllable words, such as *crime* and *color.*

Interactions Between Perception and Production

The mental node hypothesis predicts interactions between perception and production involving the same mental nodes. An example is the phenomenon of perceptual–motor adaptation, which was first reported by Cooper and Nager (1975). The subjects first listened to a synthesized acoustical stimulus resembling either /pi/ or /ti/, which was repeated continuously for about a minute over earphones. To completely eliminate muscle movement factors, the subjects held a bite board firmly in their teeth and were instructed not to mouth the sounds. After this "perceptual adaptation" phase, the subjects produced aloud the syllable /pi/ or /ti/. The dependent variable during this "test" phase was voice onset time, which was determined to be the time between the release burst of the plosive and the onset of laryngeal pulsing. The results indicated that voice onset time during *production* of /pi/ and /ti/ became systematically shorter following repeated *perception* of either /pi/ or /ti/.

The mental node hypothesis provides a simple explanation of this finding. The perception and production of segments is mediated by common components (feature nodes) that become satiated as a function of repeated activation and respond less strongly to priming. Satiation of the feature node −*voice*(voicing) during perception therefore makes it more likely that +*voice*(voicing) will become activated in error under the most-primed-wins principle during production.

W. E. Cooper, Blumstein, and Nigro (1975) obtained an effect of production on perception of approximately equal magnitude, which further strengthens this conclusion. Their subjects repeatedly produced a sequence of syllables, each beginning with a labial place of articulation: /ba ma va/. There were two conditions of articulation. Under one condition the subjects repeated the syllables aloud with normal auditory feedback, and under the other they whispered the syllables while white noise masked their auditory feedback. After repeating the syllables for a minute, the subjects identified a set of auditorily presented acoustic stimuli, which varied along the place of articulation dimension: /ba/, /da/, and /ga/. Some of the subjects showed systematic adaptation effects (with or without white noise masking their auditory feedback), and these same subjects showed an equal degree of adaptation in the (standard) perceptual adaptation task, where they *listened* to repeating speech sounds and then identified other speech sounds with varying degrees of similarity to the adaptation stimuli.

The results for these subjects indicate that speech production can influence speech perception, and this finding adds further support to the hypothesis that phonological nodes provide a common substratum underlying both perception and production. It should be noted, however, that some subjects showed no effects of either perceptual or motor adaptation on the identification of speech sounds (W. E. Cooper et al., 1975). This variability seems attributable to individual differences (some subjects appear to be especially susceptible to adaptation), standard measurement error, and the small magnitude of effects in this paradigm (even statistically reliable differences only amounted to 3 ms in some cases). Needless to say, the theoretical importance of a statistically significant and independently replicated effect is not proportional to the absolute magnitude of the effect.

It should also be noted that subsequent studies reviewed in W. E. Cooper (1979) have shown that sensory analysis nodes are also subject to adaptation.

Adaptation effects diminish by anywhere from 5% to 50% when adaptation stimuli are presented to one ear and test stimuli are presented to the other ear, indicating satiation at both binaural (mental node) and monaural (sensory analysis) sites.

Shadowing Latencies

The latencies observed in the shadowing of speech inputs further illustrate the close relationship between perception and production (Kozhevnikov & Chistovich, 1965; Porter & Lubker, 1980). In shadowing experiments, subjects hear a word or sentence, which they produce aloud with as little lag as possible. The surprising result in these studies is that some subjects can shadow with lag times as short as 100 ms between acoustic onset of input and output, even with nonsense syllables as stimuli. These shadowing latencies are faster than the fastest auditory reaction times to a pure tone stimulus (about 150 ms), using either a single-alternative key press or a single-syllable response. These short shadowing times are all the more remarkable because shadowing involves a much larger set of response alternatives, a factor normally associated with *increased* reaction time. There apparently exists an intimate relationship or direct connection between mechanisms for perceiving and producing speech (see also McLeod & Posner, 1983), and this intimate relationship is directly explained under the hypothesis that the phonological nodes for perceiving and producing speech are identical.

Speed–Accuracy Trade-off in Perception

The node structure theory was originally designed to explain the trade-off relationship between time and accuracy in motor and mental skills (D. G. MacKay, 1982), and mental nodes readily capture speed–accuracy trade-offs in perceptual recognition. To recognize an object (word), the highest level node representing the object (word) must receive greater priming than any other extraneous node in its domain when the activating mechanism is applied. Whereas the priming for extraneous nodes is unpredictable, approximating a Gausian distribution with resting level as mean, priming for the appropriate or primed-from-below node summates systematically over time and must eventually exceed the priming of every other node in its domain if the stimulus duration is sufficient. But shortening the stimulus duration increases the likelihood of error, that is, the probability that some other node will be receiving more priming than the appropriate node at the time when the activating mechanism is applied.

Specific Units for Perceiving and Producing Speech

So far we have examined evidence for the hypothesis that perception and production involve identical units above the sensory analysis and muscle movement levels. This mental node hypothesis is to some extent independent of exactly what these units are and does not rise or fall on the basis of evidence for, or against, some particular unit such as, say, the syllable. The mental node hypothesis predicts only that units playing a role in perception will also play a role in production and vice versa. With this prediction in mind, let's look briefly at the evidence for specific units, first in perception and then in production.

Studies of speech perception over the past 50 years (for a review, see H. H. Clark & Clark, 1977) call for a hierarchy of abstract units including distinctive features (e.g., unvoiced); segments (e.g., /p/); syllables (e.g., *pre*); words (e.g., *predictions*); and larger sentential constituents, such as noun phrases (e.g., *theoretical predictions*); and verb phrases (e.g., *guide research*).

Available data are consistent with the hypothesis that above the sensory analysis and muscle movement levels, perception and production involve identical units. For example, recent studies of speech errors (Fromkin, 1973) indicate that the preceding perceptual units also play a role in production. Indeed, the error data for speech production go beyond the perceptual data. Many recently discovered production units have yet to be examined in studies of auditory speech perception. Within the structure of words, these recently discovered production units include word stems, stem compounds, prefixes, and suffixes, and all of these units are specific; that is, they interact only with units of the same domain or type. For example, adverbial suffixes constitute a different type of unit from past-tense suffixes, because adverbial suffixes do not substitute in error with past-tense suffixes and vice versa (D. G. MacKay, 1979).

Within the structure of syllables, the recently discovered production units include the initial consonant group, or onset (the consonant or consonant cluster preceding the vowel); the vowel group, or rhyme (the vowel and subsequent consonants within the syllable); the final consonant group, or coda (the consonants following the vowel); the vowel nucleus (a simple vowel plus a glide and/or liquid); and the diphthong (simple vowel plus glide) (D. G. MacKay, 1979).

The mental node hypothesis predicts that all of these recently discovered production units will play a role in perception and more generally that each new abstract unit discovered in studies of production will have a counterpart in perception and vice versa. Needless to say, a great deal of additional research is needed to test this general prediction. The perceptual units that are yet to be conclusively demonstrated include word stems, stem compounds, prefixes, and suffixes, initial consonant clusters, final consonant clusters, the vowel group, and in some respects, the syllable itself.

A great deal more work also remains to be done in order to apply the mental node hypothesis to the detailed nature of phonological features. For example, several findings suggest that the phonological representations of voicing and place of articulation may differ. Both Cooper, Billings, and Cole (1976) and Meyer and Gordon (1983) observed interactions between perceiving versus producing the voicing feature, but Gordon and Meyer (1984) found no such interactions between perceiving versus producing the place of articulation feature. W. E. Cooper et al. (1976) likewise experienced difficulty using the selective adaptation technique to demonstrate interactions between perceiving versus producing place of articulation. Perhaps the sensory analysis and muscle movement nodes that represent what we now call place of articulation are connected directly with segment nodes, so that no intervening feature nodes represent place of articulation per se. Although this would explain the missing interaction, it seems too early, given our current state of knowledge, to commit a general theory on this issue.

Experimentally Induced Speech Errors

The newly discovered production units discussed above have received support from three sources: studies of naturally occurring errors (e.g., D. G. MacKay, 1972), studies of the relative ease of perceiving and producing "secret languages" resembling Pig Latin and Double Dutch (Treiman, 1983; D. G. MacKay, 1973b), and studies of experimentally induced speech errors (e.g., Baars, Motley, & MacKay, 1975). To illustrate this latter source of data, I discuss in detail the study of D. G. MacKay (1978) on experimentally induced speech errors, which not only provides data on production units such as the diphthong, but suggests an interesting means of testing the prediction that identical units play a role in perception and production.

The task was as follows: Subjects heard a series of tape-recorded syllables, presented at a rate of one every 20 s, and listened for the presence of a critical segment, either /p/ or /b/, which might or might not occur in the syllable. If the syllable contained a /p/, the subjects had to change it mentally to /b/ and produce the resulting syllable as quickly as possible. Conversely, if the syllable contained a /b/, the subjects had to change it mentally to /p/ and produce the resulting syllable as quickly as possible. For example, if the subjects heard the syllable *ban*, they said *pan* as quickly as possible, and if they heard the syllable *nip*, they said *nib* as quickly as possible. As a check for possible misperceptions, after each trial the subjects wrote down the syllable that they thought had been presented.

The original purpose of the experiment was to investigate the nature of phonological rules and to test the hypothesis that the distinctive feature *voicing* constitutes an independently controllable unit within the speech production system. Interesting evidence on both of these issues was obtained. Of interest here, however, is the fact that the subjects made hundreds of errors in both perception and production. Moreover, perception and production errors were similar and had a systematic bearing on the internal organization of phonological units within the syllable. Perceptual errors were determined from what the subjects wrote down as the syllable they perceived from the tape recording. Production errors included only the incorrect responses that occurred when the stimulus was perceived correctly and were therefore operationally independent from perceptual errors. I will first examine two classes of production errors and then discuss the implications of the perceptual errors for future research.

Diphthong Simplifications

Diphthong simplifications were studied as a class of speech errors involving complex vowels (D. G. MacKay, 1978). Under one descriptive system, that of

Chomsky and Halle (1968), complex vowels are indivisible units that might be represented E, A, I, O, U. Under another descriptive system, that of Gleason (1961), complex vowels consist of two units that might be represented /iy, ey, aey, ow, uw/. Diphthong simplifications occurred when subjects simplified a diphthong in the transformed syllable by dropping its glide, either /w/ or /y/ in Gleason's (1961) description. A typical example involved the correctly perceived stimulus /piyk/. Subjects should have said /biyk/, but frequently said /bik/, inadvertently dropping the glide, /y/. Production errors such as these suggest that somewhere in the phonological system, diphthongs consist of a simple vowel plus a glide, as in /iy, ey, aey, aw, ow, uw/. However, this is not to say that Chomsky and Halle (1968) were wrong and that diphthongs are not also indivisible units higher up in the phonological system. *Both* descriptions are correct under the node structure theory. At one level, diphthongs constitute an indivisible unit represented by a single superordinate "diphthong node." At another level, diphthongs constitute two units, represented by a vowel node and a glide node, which send bottom-up connections to and receive top-down connections from the diphthong node.

The existence of higher level phonological units, such as diphthong nodes, suggests an interesting solution to a number of unresolved controversies concerning underlying phonological representations. Examples are the debates over the divisible versus indivisible nature of affricates such as /ch/, or rhotacized vowels such as /er/, and of velar nasals such as /-ng/. The node structure theory suggests that both sides of these debates are correct. Just as words are indivisible units at one level, but not at another, affricates, rhotacized vowels, and velar nasals are indivisible units at one level but consist of separate subcomponents at another level.

Complex Vowel Substitutions

Complex vowels participated in another class of errors called complex vowel substitutions in D. G. MacKay (1978). Subjects making these errors inadvertently substituted a complex vowel for a simple vowel plus a liquid, either /l/ or /r/. An example is the substitution of *powk* for *pork*, where a glide, /w/, has replaced the liquid, /r/. Although few in number and complex in nature, errors resembling these complex vowel substitutions also occur in everyday speech production. Examples, borrowed from Fromkin (1973), are the misproduction of *soup* as *serp* and *goal* as *girl*.

Taken together, complex vowel substitutions and diphthong simplification errors suggest a new domain of vowel nucleus units, which are expressed in the surface output as a simple vowel plus either a liquid or a glide. In short, nodes in the domain (vowel nucleus) each connect with two subordinate nodes: a vowel node and one other node representing either a liquid, as in the case of rhotacized vowels, or a glide, as in the case of diphthongs.

Experimentally Induced Perceptual Errors

The mental node hypothesis predicts that production units, such as the vowel nucleus discussed above, will play a role in perception and vice versa, and the procedures of D. G. MacKay (1978) suggest an interesting technique for testing this hypothesis. Consider the misperceptions of "noncritical" consonants, that is, any consonant in the syllables except for the ones the subjects were instructed to look for, /p/ and /b/. An example is the misperception of *nip* as *mip*. When noncritical consonants were misperceived, the misperceived consonant usually differed from the actual consonant by a single distinctive feature, most often in place of articulation ($p = .90$) rather than any other feature or feature cluster ($p = .10$). More importantly, place of articulation misperceptions were highly systematic; the place of articulation of the substituting consonant was usually more frontal than the actual place of articulation of the misperceived or substituted consonant. The misperception of *napt* as *mapt* provides an example. The substituting consonant, /m/, is more frontal than the substituted consonant, /n/. This bias toward perceiving a more frontal place of articulation was highly reliable and reflected the fact that subjects were instructed to listen for /b/ and /p/, which are consonants with a frontal place of articulation. When other (control) subjects simply listened to tapes of the same syllables, and wrote down what they heard, the bias toward frontal misperceptions disappeared.

These place-of-articulation misperceptions were therefore experimentally induced, and it should be possible to induce other types of misperceptions in the same way. For example, this induction technique could be adapted to test the hypothesis that vowel nucleus units play a role in perception. If subjects are instructed to press a key as rapidly as possible to indicate the occurrence of, say, a liquid (i.e., /r/ or /l/), then, just as in production, perceptual substitution errors of liquid for glide should be common occurrences, especially when the glide is part of a diphthong, as in the substitution errors *pork* for *powk*, *serp* for *soup*, and *girl* for *goal*.

Evidence for Symmetric Connections

Having discussed some general classes of phenomena that are consistent with the mental node hypothesis, I now examine one of the implications of mental nodes, namely that some of the connections between mental nodes must be symmetric or parallel. By symmetric I mean that the bottom-up connection between two nodes has a corresponding top-down connection and vice versa. Except for the lowest level mental nodes, bottom-up and top-down connections must be symmetric whenever identical nodes are involved in perception and production. By way of illustration, the connections between corresponding nodes in Figures 2.2 and 2.3 are symmetric. The bottom-up connections in Figure 2.3 parallel the top-down connections in Figure 2.2. Symmetric connections such as these help to make sense of the otherwise puzzling production phenomena and parallels between perception and production, which follow.

Bottom-up Effects in Speech Production

Symmetric connections readily explain recent evidence for bottom-up effects in speech production. As an example of one of these effects, consider the speech errors known as blends, which occur when a speaker inadvertently combines two (or more) simultaneously appropriate words (D. G. MacKay, 1972). An example is the error *sotally*, a combination of the words *solely* and *totally* in the context "He was sotally (solely/totally) responsible for that."

The main determinants of blends are syntactic and semantic similarity. As in the above example, words that become blended belong to the same syntactic class and are virtually interchangeable in meaning within their particular context of occurrence. The seemingly straightforward explanation is that the lexical content nodes for two (or more) semantically similar words in the same domain receive precisely equal priming and become activated simultaneously under the most-primed-wins principle.

However, D. G. MacKay (1973b) and Dell (1980) showed that this top-down explanation fails to account for an additional *bottom-up* effect. Specifically, D. G. MacKay (1973b) found that words involved in blends were phonologically as well as semantically similar with greater than chance probability, and Dell (1980) showed that phonological and semantic similarity independently and reliably influence these errors. Dell (1980) also reported a parallel phenomenon for word substitutions. Like blends, substituted words are usually syntactically and semantically similar (e.g., *table* and *chair*), but some (e.g., the substitution of *pressure* for *present*) (Fromkin, 1973) are phonologically similar as well. Dell (1980) also demonstrated that this phonological similarity effect exceeded chance expectation even for syntactically and semantically similar substitutions. (See Dell, 1985a; Dell & Reich, 1980; Harley, 1984; Stemberger, 1985 for other bottom-up effects taking place during speech production.)

These findings indicate that lower level processes (the phonological representation) can influence higher level processes (the selection and misselection of which word gets produced), and such phenomena are problematic for theories postulating separate perception and production components, with strictly top-down processes for production (see also Harley, 1984). However, bottom-up effects during production are readily explained in the node structure theory (as well as similar theories such as Dell, 1985a). Because the mental nodes for perception and production are identical, the bottom-up connections required for perception automatically prime the lexical content nodes for phonologically similar words, which can then become activated in error under the most-primed-wins principle. However, these phonologically similar errors will be very rare under the node structure theory, mainly because bottom-up priming is a weak (second-order) effect. Indeed, these errors seem most likely to occur when speakers have rehearsed internally what they want to say just prior to saying it, so that higher level nodes become activated twice, in two passes as it were, first during internal speech, and subsequently during overt speech. On the initial, internal speech pass, the mental node for the correct lexical item becomes activated and then self-inhibited, providing the basis for the error that occurs on the second, overt pass. Bottom-up priming arising from the first, internal speech activation will make the lexical nodes for phonologically similar words most primed if the lexical node for the correct word is still undergoing self-inhibition. However, if, as normally occurs, overt speech is produced in a single pass, without prior internal speech, the node for the correct word will become more primed than any of the nodes receiving second-order, bottom-up priming in the same domain, greatly reducing the probability of phonologically similar errors.

Perceptually Based Production Errors

Irrelevant but simultaneously ongoing perceptual input sometimes causes errors in production, and this phenomenon is difficult to explain in theories postulating separate components for production versus perception. Meringer and Mayer (1895) and Norman (1981) compiled several naturally occurring speech errors of this type, but the Stroop effect represents a well-known experimental demonstration of the same phenomenon (Norman, 1981). Subjects in Stroop studies are presented with color names printed in several different colors of ink, and the task is to ignore the word and name the color of the ink as quickly as possible. Errors are especially frequent when the color name differs from the name of the ink (e.g., the word *green* printed in red ink), and the most common error is "data driven": the printed name (*green*) substitutes the required name describing the color of the ink (*red*).

The Stroop effect is readily explained under the node structure theory, and other similar theories, where the same mental nodes are involved in perception and production and the most primed node in a domain becomes activated automatically, regardless of its source of priming. A high-frequency word such as *green* will prime *green*(color adjective) faster and more strongly than will the visually presented color green. Because the naming of a color is a relatively rare activity, color nodes will have relatively weak (i.e., slowly transmitting) connections with their corresponding word nodes. This does not mean that Stroop interference is completely describable in "race model" terms, because priming does not automatically cause activation in the theory. However, it *does* mean that color naming will either take more time, or exhibit more errors in Stroop experiments, because in order to become activated and give rise to perception, the lexical node representing the color must achieve more priming than the lexical node representing the color name.

Top-Down Effects in Perception

Symmetric connections readily explain top-down effects in both speech and visual perception. To illustrate one such effect, consider Leeper's (1936) study, in which subjects were presented with an ambiguous figure such as Jastrow's rabbit-duck, and then answered the question, "Can you see the duck?" The subject perceived the duck and not the rabbit because the question primed (top-down) the nodes representing the visuoconceptual components of ducks. With the added bottom-up priming from the figure itself, these "duck-nodes" received the most priming and became activated under the most-primed-wins principle, thereby causing perception of the duck. The "rabbit nodes," on the other hand, only received bottom-up priming, and being less primed, did not become activated, so that the rabbit was unperceived.

Extensions of the Mental Node Hypothesis

Having outlined some general sources of evidence for the mental node hypothesis, I now argue that the hypothesis as developed so far is too simple and requires extensions along the following lines.

Semisymmetric Connections

Semisymmetric connections are one of the main reasons why the evidence discussed for symmetric connections is needed. Top-down and bottom-up connections do not *always* run in parallel, even for mental nodes. Some mental nodes have some connections that are asymmetric, for example, those that contribute a bottom-up connection but receive no corresponding top-down connection. (See Grossberg, 1982, for the contrasting claim that strictly symmetric connections are in general essential for stable cognitive coding, and see Rumelhart, McClelland, & the PDP Research Group, 1986, for some models that only incorporate symmetric connections.) By way of illustrating these asymmetric connections, consider again the McGurk effect, the fact that seeing someone produce a speech sound can influence how the auditorily presented sound is perceived. Visual features such as lip closure exert a strong effect on what phoneme subjects report hearing when they see the lip movements for one syllable while hearing the sound of a different syllable (McGurk & MacDonald, 1976). Presented with a visual /pa/ and an auditory /ta/, for example, subjects usually report hearing the /pa/. Nodes representing visual lip movements apparently connect bottom-up with phonological nodes, so as to influence which phoneme node receives most priming and becomes activated.

However, there are neither logical nor empirical grounds for postulating a symmetric top-down connection between phonological nodes and the visual nodes representing lip movements. For example, hearing a speech sound over the telephone doesn't normally cause or even enable one to visualize how its production might *look*. This suggests an asymmetry. Visual nodes representing lip movements send bottom-up connections to phonological nodes but receive no top-down connections in return.

The lowest level mental nodes in an action hierarchy always have semisymmetric connections. By way of illustration, phonological feature nodes have semisymmetric connections. Connections with higher level phonological nodes are symmetric, but connections with lower level (muscle movement and sensory analysis nodes) are asymmetric. For example, phonological feature nodes send top-down connections to the muscle movement nodes for articulating speech but receive no corresponding bottom-up connections in return.

Other, higher level nodes may also have semisymmetric connections. Some mental nodes that are necessary for producing a behavioral sequence may lack corresponding bottom-up connections for reasons of structural economy and speed of processing. Consider monosyllabic words such as *desk*, for example. Producing this word requires both a lexical node, *desk*(noun), and a syllable node, *desk*(stressed syllable). Without the syllable node for monosyllabic words, speakers would be unable to produce the rhythmic timing characteristics of English (Chapter 5). However, the syllable node, *desk*(stressed syllable), may be unnecessary in perception, and may even slow down the perceptual process. That is, in monosyllabic words, phonological units such as *d*(initial consonant) and *esk*(vowel group) may connect with their lexical node directly rather than indirectly via a syllable node such as *desk*(stressed syllable). This would speed up perceptual processing but would introduce asymmetric connections at relatively high levels in the network. Testing for such high-level asymmetries is an important area for further research.

Kinesthetic and Muscle Spindle Inputs

So far I have discussed mental and muscle movement nodes as if they formed an either–or dichotomy. I represented muscle movement nodes as having no sensory or perceptual functions whatsoever. This representation is only partly correct. Although the distinction between mental and muscle movement nodes is functionally important, an analysis of kinesthetic and muscle spindle inputs suggests that this sensory–motor dichotomy is too simple. Even the very lowest level muscle movement nodes, which connect with the muscles themselves, receive direct connections from *some* sensory nodes. Specifically, sensory fibers located in spindles within the muscles connect with the lowest level alpha motorneurons, which move the muscles.

Kinesthetic feedback returns to muscle movement nodes at an only slightly higher level, perhaps still in the spinal cord. Kinesthetic input is anatomically specific and cannot be considered to connect directly with even the lowest level mental nodes representing, in the case of speech, distinctive features or phonemes. Rather, kinesthetic inputs must connect with and prime higher level muscle movement nodes.

This analysis suggests that muscle movement nodes make up a modality consisting of several hierarchically organized systems. Moreover, these muscle movement systems must themselves consist of subsystems that can be independently activated. In speech production, for example, we can activate the supralaryngeal subsystem independently from the laryngeal subsystem, which enables us to whisper, producing the same articulatory gestures but devoicing all of our speech sounds. Or we can activate the supralaryngeal articulatory subsystem independently from all other subsystems, producing lip, tongue, and jaw movements without any sound, the so-called mouthing of speech sounds.

Nature and Degree of Sensory–Perceptual Connectivity

If, as the preceding discussion suggests, the number and nature of inputs from the sensory–perceptual–cognitive systems provide the primary basis for distinguishing between mental versus muscle movement nodes, different systems of mental nodes can be distinguished in the same way. By way of illustration, compare the connections to phonological versus sentential system nodes. In particular, compare the connectivity of *p*(initial consonant) versus *pear*(noun). *P*(initial consonant) receives two possible sources of relatively direct input: from acoustic analysis nodes representing the phoneme and from visual nodes representing the lip movements. The sentential node *pear*(noun) on the other hand receives five possible sources of relatively direct input: from phonological nodes representing the word, including *p*(initial consonant); from visual concept nodes representing the visual form of a pear; from orthographic nodes representing the word *pear*; and finally, from olfactory and gustatory representations, because pears can be recognized and named from their smell and from their taste. This example further illustrates the nested nature of modalities. The visual modality contains at least three other modalities for representing visual lip movements, orthography, and visual form. Note, however, that different lexical concept nodes will receive different types of sensory–perceptual–cognitive input. For example, *dog*(noun) must receive an additional source of input from the auditory concept system, because dogs can be recognized from the sound of their bark. In general, then, nodes in higher level systems receive many connections from a variety of high- and low-level systems, whereas nodes in lower level systems receive fewer connections and mostly from low-level rather than high-level systems.

3
The Sequencing of Action

> Not only speech, but all skilled acts seem to involve the same problems of serial ordering, even down to the temporal coordination of muscular movements in such a movement as reaching and grasping. Analysis of the nervous mechanisms underlying order in the more primitive acts may contribute ultimately to the solution of even the physiology of logic. . . . Serial order is typical of the problems raised by cerebral activity; few, if any, of the problems are simpler or promise easier solution. We can, perhaps, postpone the fatal day when we must face them, by saying that they are too complex for present analysis, but there is a danger here of constructing a false picture of those processes we believe to be simpler.
>
> (Lashley, 1951, pp. 122, 197)

Any theory of action must deal with three basic questions: What is the structure of the components representing skilled behavior? How are these components activated in proper sequence? And how are these components timed or produced at the appropriate rate? The previous chapter examined the first of these problems, the structure of the components for organizing everyday actions, and the present chapter examines the second, how these components become activated in proper sequence.

I begin by outlining the general requirements for a theory of sequencing in action. I then develop a theory that meets these general requirements and makes new predictions for future test. In Chapter 4, which deals with perceptual processes, I examine the related problem of sequencing in perception.

The Sequencing of Action

How do we execute sequences of behavior in proper serial order when we do and in improper order when we make errors? As Lashley (1951) pointed out, sequencing is a general problem for psychological theories. Any behavior more complex than a spinal reflex is sequentially organized and requires explanation in a general theory of sequencing. However, speech provides the most extensively studied example of the sequencing problem. Other cerebral activities may employ similar sequencing mechanisms (Lashley, 1951; D. G. MacKay, 1985), but sequencing is especially complex and interesting in the case of speech production because sequencing issues arise at many different levels at once. In discourse, how do we produce sentences one after the other in logical order? How do we order the words within the sentences? How do we order the morphemes, syllables, and segments that make up the words? Finally, how do we order the muscle movements that give rise to the sequence of sounds? Our everyday capacity to organize and produce such a hierarchy of simultaneous, nested sequences is probably fundamental to our uniquely human ability to use language (see also Keele, 1987).

Besides being a multilevel and omnipresent issue, sequence also plays an essential role in all languages. How the components at any given level are sequenced makes a fundamental difference to the significance of an utterance. Whatever the language, changing the order of phonemes in a word, for example, changes the meaning of the word. Neither the omnipresence nor the significance of sequence is unique to language, however; both characteristics are apparent in many other everyday activities. Consider the normal sequence of actions required to light a candle, for example: (1) light the match, (2) apply the match to the wick of the candle, and (3) blow out the match. Performing these actions in any other order, such as (1) light the match, (2) blow out the match, and (3) apply the match to the wick of the candle, is analogous to producing a nonsense word. The overall behavior changes in significance when its component actions are performed in this new order.

General Requirements for a Theory of Sequencing

Any theory of sequencing must address a relatively small number of fundamental questions: Is there a nonsequential or preparatory stage that precedes the sequential activation of behavior? What is the relationship between the sequencing mechanism and the output units for producing behavior? Can sequencing be accomplished by mechanisms responsible for timing? What is the relationship between mechanisms for sequencing and timing in behavior? As we will see, available data bearing on these questions impose general constraints on all viable theories of sequencing.

Preparation for Sequencing

Lashley (1951) was the first to recognize that a priming or preparation stage is necessary for sequencing. According to Lashley, a set of output units must be primed or simultaneously readied for activation before an independently stored sequencing mechanism can activate and impose order on them. Lashley (1951) outlined three sources of support for his idea that simultaneous priming precedes sequential activation. One was anticipatory errors in speech, where an upcoming or soon-to-be-produced word or speech sound becomes produced before its time. An example is, "We have a laboratory, I mean, computer in our own laboratory." Anticipatory errors are the most frequent general class of speech errors and indicate that prior to actual activation soon-to-be-produced units are simultaneously preexcited, primed, or readied for activation. Otherwise, why would an upcoming or about-to-be-produced unit be so much more likely to intrude than any other unit in the speaker's vocabulary?

Another argument for a (simultaneous) preparatory stage prior to (sequential) activation is that "a general facilitation, a rise in the dynamic level" seems necessary for the performance of many sequential activities (Lashley, 1951, p. 187). For example, when sufficiently aroused, brain-damaged patients can execute sequences of behavior that under normal circumstances they cannot. An aphasic who is unable to produce the word "watch" in a laboratory test may exclaim, "Give me my watch!" when someone pretends to make off with his or her watch (H. L. Teuber, personal communication, April, 1965). Such examples suggest that an output sequence cannot become activated unless its units become sufficiently primed. Of course, motivational factors contribute to the required level of priming in this particular example from neuropsychology, whereas in general, priming normally arises mainly from factors specific to the action being produced.

Lashley also noted evidence for priming in studies of reaction time and word association.

> Reaction time, in general, is reduced by preliminary warning or by instructions which allow the subject to prepare for the specific act required. In controlled association experiments, the subject is instructed to respond to the stimulus word by a word having a certain type of relation to it, such as the opposite or a part of which the stimulus is the whole: black–white, apple–seed. The result is an attitude or set which causes the particular category to dominate the associative reaction. (1951, p. 187)

It is as if controlled association instructions simultaneously prime or ready for activation an entire category of specific responses, thereby short-circuiting the first stage of the prime-then-activate process, so that the appropriate response can be produced soon after presentation of the stimulus.

Lashley's (1951) third basis for assuming that priming precedes sequential activation during production is that perception exhibits a similar process. To demonstrate perceptual priming, Lashley auditorily presented to his audience the garden path sentence, "Rapid righting [writing] with his uninjured hand saved from loss the contents of the capsized canoe." As might be expected, a sudden reinterpretation of the word *writing (righting)* took place once the audience heard the last two words of the sentence (see also Carroll, 1986). On the basis of this demonstration, Lashley argued that the units for comprehending the concept "righting" (rather than "writing") could not become activated until the phrase "capsized canoe" had occurred and so must have been held in a state of readiness or partial activation "for at least 3 to 5 seconds after hearing the word" (1951, p. 193). Thus, priming, or readying for activation, precedes actual activation during comprehension, and by analogy during production as well, because "the processes of comprehension and production of speech have too much in common to depend on wholly different mechanisms" (1951, p. 186).

Independence of Sequence and Content

The mechanism for sequencing behavior must be separate from the units that represent the content or form of the behavioral sequence; the basic units representing perception and action must be independent of their sequencing mechanism. To see why this is so, consider a class of theories where sequencing and content are nonindependent: chain-association theories. There is no content-independent sequencing mechanism in chain-association theories. Unidirectional links between content nodes provide the representation of sequence. Activating the first content node directly causes activation of the second (connected) content node, and so on, until the entire sequence has been produced.

Many variants of this unidirectional bond assumption have been proposed, and the bonds are usually excitatory in nature. But not always. For example, Estes (1972) proposed a chain-association theory where the bonds are inhibitory rather than excitatory. The first unit inhibits the remaining units, the second inhibits all but the first, the third inhibits all but the first two, and so on. For example, in producing a simple word such as *act*, a superordinate node representing the entire word becomes activated and primes its three subordinate nodes representing the segments, /a/, /c/, and /t/. Now under the unidirectional bond assumption, the first element, representing /a/, inhibits the other two, and the second element, representing /c/, inhibits the third, representing /t/. Thus, the first element, not being inhibited by any of the others, achieves the greatest degree of priming, and becomes activated. The second, no longer being inhibited by the first, now has the greatest priming and becomes activated, releasing the third from inhibition, and so on.

Lashley (1951) pointed out the basic problem with this and related chain-association proposals. The problem is that links between the basic output components will interfere with one another. For example, either excitatory or inhibitory links between the components for *act* will prevent error-free production of *cat* and *tack*, or any other words containing the same components in a different order and vice versa. Extrapolating to a normal 50,000-word vocabulary, conflicting connections between the basic output components would simply prevent speech altogether. Of course, it might be suggested that sequential connections between content nodes are not permanent but are established on the spot as part of the preparation for sequencing (Norman & Rumelhart, 1983). However, this suggestion simply begs the question and adds a new unresolved issue: How are the appropriate (and no other) connections formed, and how are they formed so quickly?

Theories postulating nonindependent sequencing and content units therefore fail to explain the production of sequence per se. These theories also predict errors that do not occur and have difficulty explaining the errors that do occur (D. G. MacKay, 1970e). Because they postulate nonindependent mechanisms for sequence and content, these theories also have difficulty explaining flexibility in sequential behavior. Children's word games, for example, Pig Latin, illustrate the nature of this flexibility (D. G. MacKay, 1972). When playing Pig Latin, children quickly and easily impose a new order on the segments of both never previously encountered nonsense syllables, such as *snark*, and frequently used words, such as *pig* (see also Treiman, 1983). When children produce the word *pig* as *igpay*, for example, no painful process of unlearning the old habitual sequence is required, as might be expected if the old sequence were built into the output units themselves by means of undirectional links. Instead, the sequencing mechanism appears to operate on the basis of internal rules that can be easily altered and that can apply to an indefinitely large number of behavioral units, including never previously encountered ones such as *snark*.

Lashley (1951) noted one final set of phenomena calling for a rulelike sequencing mechanism that is independent of the content units themselves: the ability to translate freely from one language to another using different word orders. An experienced translator does not have to proceed word by word, but can and often must rapidly alter the order of the components making up the original idea. Such flexibility suggests that sequence is not part of the ideas per se, but is imposed on these ideas by a language-specific sequencing mechanism. The way that bilinguals sometimes impose the *wrong* order on words likewise suggests that the sequencing mechanism is independent of the words and ideas being sequenced. A native speaker of German may sometimes impose aspects of German syntax when speaking English, postponing the verb to the end of a familiar English expression, for example. One might argue about whether and how the sequencing instructions for one language can become attached to the word units for the other language in these examples, but such errors simply could not occur if sequencing mechanism and word units were inseparable.

Sequencing and the Initiation of Behavior

Theories of sequencing must explain a special and repeatedly demonstrated relationship between sequencing and the initiation of behavior. Studies such as Klapp et al. (1973), Sternberg et al. (1978), and Klapp and Wyatt (1976) have shown that it takes less time to initiate a preplanned behavior that consists of a single component than one that consists of a sequence of components. This relationship between sequencing and the initiation of behavior is an embarrassment to chain-association or horizontal link theories, even those augmented with vertical links (e.g., Estes, 1972; Wickelgren, 1979). It also presents problems for theories incorporating a scanning mechanism. In scanning theories, such as D. G. MacKay (1971), a behavioral sequence is loaded into a memory buffer in preparation for sequencing, and behavior becomes initiated by a scanner that sweeps over the buffer from, say, left to right. Thus, a subject who is prepared to say the word *paper*, for example, has already loaded *paper* into the output buffer, and following a go or "speak now" signal, the scanner sweeps over the buffer, causing activation of the initial /p/, followed in turn by the remaining segments of the word. This sequencing process is of course independent of word length, so that the scanner should trigger the initial /p/ of a one-syllable word such as *paint* no faster than the initial /p/ of a two-syllable word such as *paper*.

Available data do not support this prediction, however. As noted in the previous chapter, production onset time is significantly longer for two-syllable words such as *paper* than for one-syllable words such as *paint* (Klapp et al., 1973, among others). Klapp and Wyatt (1976) also observed a similar relationship between sequencing and the initiation behavior in production onset times for sequences of finger movement. Subjects in Klapp and Wyatt (1976) produced one of four Morse code sequences on a telegraph key: *dit-dit*, *dit-dah*, *dah-dit*, and *dah-dah* (to produce a *dit*, the key is released immediately after the press, and to produce a *dah*, the key is held down for about 200 ms prior to release). A light indicated which of the four response sequences to produce, dependent variables being production onset time and the time between the first and second response components. The subjects were of course college students with no prior experience in generating Morse code.

There were three main results: (1) Production onset time for sequences beginning with *dit* was shorter than for sequences beginning with *dah*, but the nature of the second response (*dit* versus *dah*) had no effect on production onset time. (2) The time to initiate the second response (following the first) was longer for *dah* than for *dit*. (3) Production onset time was much faster, however, when the second component was identical to the first (both *dits* or both *dahs*) than when the second was different (e.g., a *dit* and a *dah*).

To explain these results, Klapp and Wyatt (1976) reasoned that planning a *dit* was simpler than planning a *dah*, that only the first response was planned during production onset time, and that the second response was planned during the interresponse interval following the first. However, the third observation contradicted this explanation. Because production onset time was shorter when the second response was the same as the first, the second response must have been planned prior to initiating the first. This seemingly contradictory finding has remained unexplained since Klapp and Wyatt's (1976) study.

Errors in Sequencing

Theories of sequencing must of course explain how sequential errors occur—not just the fact that sequential errors occur, but the detailed nature of the regularities that have been observed in these errors. An example is the sequential class phenomenon, one of the strongest and most general regularities observed to date in speech errors. The phenomenon is this: When a speaker inadvertently substitutes one linguistic component for another, both components usually belong to the same sequential class. Cohen (1967) originally observed this regularity in errors involving interchanged words. An example is the error, "We have a laboratory in our own computer," where one noun (*laboratory*) interchanges with another (*computer*). As in this example, nouns generally interchange with other nouns, verbs with verbs and not with, say, nouns or adjectives (Cohen, 1967). Even "Freudian slips" such as, say, "He found her crotch, I mean, watch" (example modified from Fromkin, 1973), adhere to this sequential class rule. Because both *watch* and *crotch* are nouns, this example obeys the sequential class regularity, even though, as Fromkin (1973) points out, semantic (Freudian) factors may also have played a role.

The sequential class regularity has also been observed for errors involving the following components: (1) Morphological components: prefixes interchange with other prefixes, suffixes with other suffixes, and never prefixes with suffixes (D. G. MacKay, 1979). (2) Syllabic components: initial consonant clusters interchange with other initial clusters, and final with final, but never initial with final (D. G. MacKay, 1972; semivowels of course excluded, see Stemberger, 1983). (3) Segmental components: vowels interchange with vowels, consonants with consonants, and never vowels with consonants (D. G. MacKay, 1972). In short, the sequential class regularity holds for all levels of speech production, and a viable theory of sequencing must explain this fact.

Practice Effects

Why do some behaviors exhibit sequencing errors, but not others? For example, humans don't make purely sequential errors in walking and neither do horses. Similarly, we almost never make sequential errors involving the phonological components of function and content words, producing *ce that* for *the cat*, for example. Frequency/practice is almost certainly responsible for this function word effect (see the recent experiments of Dell, 1985b) and may contribute to the absence of sequential errors in walking as well (D. G. MacKay, 1982).

Different Mechanisms for Sequencing and Timing

Another general constraint on theories of sequencing is that different mechanisms are required to time and to sequence behavior. Sequencing cannot be achieved by a timing mechanism, and timing cannot be achieved by a sequencing mechanism. In what follows, I examine both of these hypothetical possibilities in turn to show why neither works.

Consider first the possibility (proposed by Rosenbaum, 1985) that a timing mechanism is by itself responsible for both sequencing and timing in speech production. This hypothetical timing mechanism is able to generate the sequence of phonemes in a word by specifying their time of production, and sequencing errors arise because phonemes have been assigned improper times. For example, the word *cat* might be misproduced as *act* because the *a* has been produced relatively early and the *c* produced relatively late. Likewise, at a higher level, the phrase "in the car," might be misproduced as "in car the," because the noun is produced relatively early, and the article relatively late.

The problem with this account is that in general, substituted components in actually occurring sequential errors don't just exchange places in time; sequential class almost invariably plays a role. For example, in the error "cake the ring of teas" instead of "take the ring of keys," the segments /*t*/ and /*k*/ exchange temporal positions, but they also belong to the same domain or sequential class, *initial consonant group*. Even in haplologies or "skipping errors"—such as *shrimp and egg* misproduced as *shrigg*, skipping *—mp and e—* (from Stemberger, 1985)—the speaker skips to a component in the same sequential class (final consonant group in this example) as was required for the intended word. These and other sequential regularities (Stemberger, 1985) would not be expected if a timing mechanism determines sequencing.

Consider now the other hypothetical possibility, that a sequencing mechanism determines both sequencing and timing, an idea proposed by Norman and Rumelhart (1983), among others. Norman and Rumelhart's (1983) theory of typing incorporates a sequencing mechanism, but no timing mechanism, and timing depends on the nature of the operations required for sequencing. Under this view, errors in the timing and sequencing of typestrokes are one and the same. When typestrokes occur out of sequence, one component is being activated especially early, and the other is being activated especially late.

A critical piece of data contradicting this hypothesis appears in Grudin (1981). Grudin had skilled typists type a large corpus of text, and examined their keystroke intervals, the time between one keystroke and the next. He was especially interested in the keystroke intervals for inadvertently produced transposition errors, where *the* is mistyped as *hte*, for example. The results, averaged over a large number of two-letter transpositions, showed no tendency for one key to come especially early and the other especially late. Rather, the keys exchanged places both in sequence, and in time, just as in speech errors. For example, assume that a skilled typist normally types the word *the* correctly with about 140 ms between hitting *space* and *t* and 75 ms between hitting *t* and *h*. Grudin found that when this typist produced the transposition error *hte*, timing remained the same, about 140 ms between space and *h* and 75 ms between *h* and *t*. The wrong components were activated at the right time. This finding indicates that timing is independent of the behavior being timed, and this independence could only occur with separate mechanisms for determining the content, sequencing, and timing of behavior.

Grudin's (1981) findings also indicate that timing is being "programmed" in proficient typing, and this is an especially important fact for theories of sequencing and timing, because typing is a skill that does not demand consistent or accurate timing, unlike, say, music or Morse code. Apparently a timing mechanism plays a role in skilled behavior even when precise and consistent timing is unnecessary.

Close Relationship Between Sequence and Timing

A final requirement for theories of sequencing is a close relationship between mechanisms for sequencing and mechanisms for timing; even though timing and sequencing mechanisms are independent of one another, they must nevertheless be closely connected. Findings of Schmidt (1980) can be interpreted as providing preliminary evidence for this close connection. Schmidt had subjects practice moving a lever to a target in a specified period of time, giving them feedback on their movement time following each trial. Different subjects practiced two different types of movement: one was sequential, the other nonsequential. In the nonsequential condition, subjects simply moved the lever horizontally to the specified target in the specified amount of time. In the sequential condition, two movements were required: first to the target and then back to the start position, again in a specified period of time.

The independent variable was a change in the mass of the lever. Suddenly, and without warning, the lever became more difficult to move. This change had strikingly different effects on sequential versus nonsequential movements. In completing the nonsequential movement with the sluggish lever, the subjects reached the target but took longer to do so. However, in completing the sequential movement, they finished at the correct time but undershot the target in space. When producing highly practiced *sequential* movements, we program the time to produce the components of the sequence in advance, and we find it difficult to change this preprogrammed timing, as if sequence and timing involve closely coupled mechanisms.

The Node Structure Theory of Sequencing

In addition to explaining the constraints discussed above, a node structure theory must also explain how nodes with the dynamic properties (e.g., priming, activation, and linkage strength) and the structure of interconnections discussed in previous chapters give rise to the sequential organization of rapidly produced actions. As others have noted, this is a major problem for parallel distributed processing (PDP) theories (see McClelland et al., 1986).

The fact that content nodes are hierarchically organized in the node structure theory means that more nodes must be activated in sequence than if content nodes were organized on only a single level. As we have seen, however, sequencing which is applied at a single level cannot explain natural skills such as speech production. Moreover, the benefits of hierarchical organization greatly outweigh the costs in additional sequence nodes and sequential decisions. Hierarchies facilitate creativity and flexibility of expression; different lower level expressions of the same higher level content become possible in a hierarchic skill. As Keele (1985) and others have pointed out, hierarchies also cut down on how much new learning or connection formation is required. Preformed lower level node structures simply become attached to new higher level nodes and used for new purposes. For example, no new learning is required at the phonological level when adults encounter a new word formed by reassembling morphological components of already familiar words (see also D. G. MacKay, 1982).

The theory I develop here straightforwardly extends the theory I proposed (D. G. MacKay, 1982) for explaining how practice makes behavior more fluent (faster, less prone to error) and more flexible (adapting readily to changed circumstances and transferring readily from one response mechanism to another). Only minor modifications have been necessary for the purpose of developing a unified theory of perception and action.

Sequencing in Action Hierarchies

The content nodes in Figure 3.1 illustrate the problems of sequencing and nonsequencing as they apply to the node structure theory. Some of the nodes in Figure 3.1 must be activated simultaneously rather than sequentially. Distinctive feature nodes receiving simultaneous priming from the same segment node represent an example. All of the distinctive feature nodes for the /p/ in *prove* must

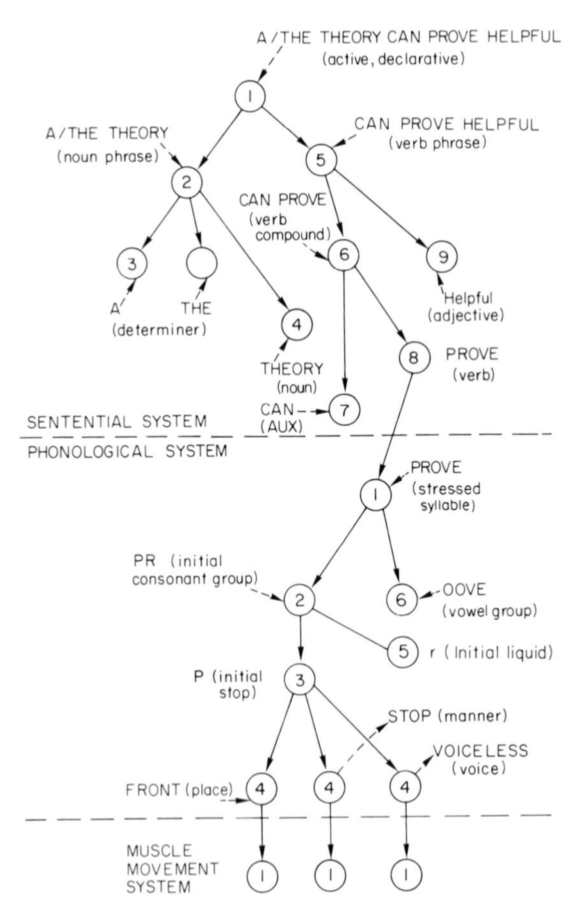

FIGURE 3.1. Aspects of the action hierarchy for producing the sentence "A/the theory can prove helpful." Note that node activation is sequenced within a system, so that the numbers within nodes denote relative order of activation within their respective systems.

become activated at the same or nearly the same time. Achieving this simultaneity constitutes a serial-order problem in reverse.

However, most of the nodes in Figure 3.1 must be activated in sequence if the output is to be error free, and the numbers within the relevant nodes represent the order of activation within a system. The highest level node, *a/the theory can prove helpful*(active declarative), must be activated first. This simultaneously primes both *a/the theory*(noun phrase) and *can prove helpful*(verb phrase), but only *a/the theory*(noun phrase) must become activated at this point. Activating this node simultaneously primes its connected nodes, *a*(determiner), *the*(determiner), and *theory*(noun).

Similarly, activating *prove*(verb) primes a syllable node in the phonological system, which in turn primes *pr*(initial consonant group). *Activating pr*(initial consonant group) primes two connected nodes representing the segments *p* and *r*. Activating *p*(initial consonant) primes a set of phonological feature nodes, including one representing the frontal place of articulation of *p*. Finally, activating the frontal feature node primes a set of muscle movement nodes, including one for contracting the obicularis oral muscles of the lips. Full-fledged behavior of course only takes place if, and only if, the muscle movement nodes at the lowest level of the action hierarchy become activated. The issue, then, is this: How are all these nodes activated in proper order?

The Sequencing Mechanism

So far, the book has focused mainly on the basic components of the theory, content nodes for organizing the form or content of a preplanned action. I have explained why content nodes are hierarchically organized and why they each represent a class of actions, so that the content node representing, say, the phoneme /p/, controls all of the context-dependent ways of pronouncing a /p/, including whispering and shouting. I now elaborate on sequence nodes, their rationale, the structure of their connections, their relationship to content nodes, and the differences between sequence and content nodes in the theory. I then apply the node structure theory of sequencing to specific types of data and outline some predictions of the theory for future test.

Sequence nodes are the triggering mechanisms that determine whether, and in what order, the content nodes in an action hierarchy become activated. The sequencing process involves categories or domains of content nodes, rather than individual content nodes, as in chain-association theories. Recall that a domain consists of all of the nodes connected to a single sequence node. Stated differently, all of the nodes in a domain become activated by means of the same triggering mechanism. The sequence code COLOR ADJECTIVE, for example, is connected to, and serves to activate the many content nodes in the domain *color adjective*. When activated, a sequence node multiplies the priming of every node connected with it by some large factor within a relatively brief period of time. (For discussions of multiplicative gating processes and their possible physiological implementations, see Grossberg, 1982; Sejnowski, 1981.) This multiplicative process has no effect on an unprimed node but soon serves to activate (i.e., bring to threshold) the content node with the greatest degree of priming in its domain, normally the one that has just been primed from above via a connection from a superordinate content node. In producing the adjective *green*, for example, *green*(color adjective) must first become primed, either from above, via a superordinate node such as *green apples*(noun phrase), or from below, via, say, visual perception of either the color green or the printed word *green*. A content node receiving first-order priming passes second-order priming to its connected sequence node, in this case, COLOR ADJECTIVE. Then the sequence node becomes activated (by a process discussed in the next section) and in turn activates the most primed content node in its domain, in this case, *green*(color adjective). This most-primed-wins principle is extremely general and governs the activation of sequence nodes as well as content nodes (D. G. MacKay, 1982).

Without becoming quenched, a sequence node can self-sustain its activation for a set period, say 20 ms, and multiplies the priming of its connected nodes by, say, a factor of 2 every millisecond. The original degree of priming for a content node must therefore exceed some minimal level, so that multiplication of priming by the sequence node over 20 ms can achieve the threshold level required for self-sustained activation of the content node. If the threshold level is 100, this minimal level is of course 100 divided by 2^{20} in this example. Below this minimal level, the multiplied priming of the content node remains subthreshold, and activation cannot occur. And if a node has an initial level of priming of 10, multiplication of priming will reach threshold (100) in less than 4 ms.

Sequential Rules

Connections between sequence nodes represent serial-order rules that determine order of activation when two or more sequence nodes have received simultaneous priming. Serial-order rules represented by sequence nodes in the sentential system are termed syntactic rules, and serial-order rules represented by sequence nodes in the phonological system are termed phonological rules (see Table 3.1 for typical examples). The sequence nodes ADJECTIVE and NOUN, for example, are connected in such a way as to represent the syntactic rule that adjectives precede nouns in English noun phrases. Similarly, the sequence nodes INITIAL CONSONANT GROUP and VOWEL GROUP are connected in such a way as to represent the phonological rule that initial consonants in a syllable precede the vowel and final consonants.

An inhibitory connection is a simple means of achieving this order relation among sequence nodes. Under this proposal, the first to be activated of a pair of sequentially organized sequence nodes inhibits the next to be activated whenever both are simultaneously primed. This inhibitory connection therefore enables the first to be activated to become activated first under the most-primed-wins principle. Following activation of the first sequence node, the next to be activated is released from inhibition and can then become activated under the most-primed-wins principle. For example, ADJECTIVE inhibits NOUN and dominates

TABLE 3.1. Examples of serial-order rules in the phonological and sentential systems.

Example Serial-Order Rules	Example Instances
Phonological System	
Initial consonant group + vowel group	*str* + *and*
Initial fricative + initial stop + initial liquid	*s* + *t* + *r*
Vowel + glide + liquid	*o* + *w* + *l*
Final nasal + final stop	*n* + *d*
Sentential System	
Noun phrase + verb phrase	*the theory* + *enabled extensive progress*
Verb + noun phrase	*enabled* + *extensive progress*
Determiner + noun	*the* + *theory*
Adjective + noun	*extensive* + *progress*

in degree of priming whenever ADJECTIVE and NOUN receive simultaneous priming. However, once ADJECTIVE has been activated and its priming returns to resting level, NOUN is released from inhibition and dominates in degree of priming, thereby determining the sequence (adjective + noun) for this and any other noun phrase containing an adjective and a noun.

Sequential rules such as adjective + noun bear a surface resemblance to phrase structure rules of Chomsky (1957), such as noun phrase ⟶ noun + adjective, where the arrow stands for "is rewritten as." Both types of rules are nontransformational, for example, and refer in this example to identical sentential domains or syntactic categories, the set of all adjectives and nouns. There are many differences, however (D. G. MacKay, 1974). For example, there is no sense in which the sequence node NOUN PHRASE is "rewritten" as NOUN + ADJECTIVE in the node structure theory. Rather, the lexical content nodes connected to this particular noun phrase node simultaneously prime their respective sequence nodes, which happen to be NOUN and ADJECTIVE.

The node structure theory also postulates new processes and sequential domains that were unforeseen in phrase structure grammars, such as sequential domains for discourse nodes (D. G. MacKay, 1985), for morphological nodes (D. G. MacKay, 1973b), and for phonological nodes, as in *initial consonant group + vowel group*. Moreover, the seeming equivalence of some phrase structure categories and sequential domains in the examples discussed so far is fortuitous. Sequential domains such as *adjective* and *noun* are in fact much more fine grained than I have so far discussed. For example, the domain of *green* and *red* must be *color adjective*, and the domain of *frequent* and *fast* must be *temporal adjective*, rather than just *adjective*. Although all adjectives precede nouns in English, more restricted domains, such as *color adjective* and *temporal adjective*, are necessary in order to ensure the appropriate sequencing among different types of adjectives. Thus, a sequential rule such as *temporal adjective + color adjective* is required in order to produce the usual sequence when temporal and color adjectives are conjoined. For example, we normally say "frequent red lights," instead of "red frequent lights." Without domains such as *color adjective* and *temporal adjective*, we would have no mechanism for producing the preferred rather than nonpreferred adjective order. Factors such as emphasis can of course alter this preferred or neutral order.

Subdivisions within the domain that I have represented here as nouns are likewise necessary, and for similar reasons. The only constraint on sequential domains in the theory is that their corresponding sequence nodes must be "called up" by means of direct connections from content nodes. Is it possible to obey this constraint in describing the overall system of sequential rules for English or any other language? What is the full set of sequential domains for producing English? These questions currently lack conclusive answers (but see Gazdar, 1981).

However, the way that content nodes call up sequence nodes in parallel and the way that sequence nodes, after interacting among themselves, activate content nodes seems to be exactly the sort of mutually interacting process that is needed for explaining recent demonstrations of an influence of syntax on word recognition (Isenberg, Walker, Ryder, & Schweikert, 1980) and on word selection (Bock & Warren, 1985) and vice versa (Bock & Warren, 1985).

The Activation of Sequence Nodes

Timing nodes are the mechanism for activating sequence nodes and are connected with sequence nodes in the same one-to-many way that sequence nodes are connected with content nodes. Separate mechanisms therefore determine the form, sequence, and timing of behavior in the node structure theory, but timing is more closely related to sequencing than to the form of behavior; timing nodes are directly connected to sequence nodes, but not to content nodes.

Timing nodes become activated according to an endogenous rhythm, and timing nodes at different levels have different endogenous rhythms (Chapter 5). Following each activation, timing nodes multiply the priming of the sequence nodes connected to them, activating the most primed one on the basis of the most-primed-wins principle. By determining how rapidly the sequence nodes become activated, timing nodes therefore determine the rate of the output, a topic taken up in detail in Chapter 5, along with other aspects of the temporal organization of perception and action.

Functional Relationships Between Sequence and Content Nodes

Sequence nodes perform three main functions in relation to content nodes: They organize the content nodes into domains, they activate the most primed content node in a domain, and they determine the serial order in which the content nodes become activated.

The Organizing Function

A domain consists of a set of nodes that all become activated by means of the same mechanism. For example, the hypothetical sequence node TEMPORAL ADJECTIVE is connected to, and serves to activate, the many content nodes in the domain *temporal adjective*, thereby organizing these nodes together into a single domain. However, domains reflect a functional relationship among nodes. Domains should not *in general* be thought of as sets of nodes aggregated into nonoverlapping anatomical areas of the brain. Although such localization may actually occur in lower level systems (D. G. MacKay, 1985), one and the same content node in a higher level system can receive connections from several sequence nodes and thereby occupy several domains. For example, consider the sentential system representation of the many English words that can be used with identical meaning as either nouns or verbs, for example, *practice*. A single content node, *practice*(noun, verb), represents this word and becomes activated by either NOUN or VERB. Because dual-function content nodes, such as this one, simultaneously occupy more than one domain, higher level domains must overlap, at least to some extent, at the neuroanatomical level.

From a functional point of view, a sequential domain consists of a set of nodes, all of which have the same sequential function. Thus, the domain of content nodes representing nouns all have the same sequential properties or privileges of occurrence in English sentences. Similarly, the domain of nodes representing vowels all have the same sequential properties, or privileges of occurrence, in the syllables of English or any other language.

The Triggering Function

The second function of sequence nodes is to activate whichever content node has greatest priming in its domain. This most-primed-wins principle for activation follows directly from the nature of connections between sequence and content nodes. Once a sequence node becomes activated, it automatically and simultaneously multiplies the priming of the entire domain of content nodes connected with it, so that their level of priming increases rapidly over time. However, the intended-to-be-activated node in the domain has just received priming "from above," because its superordinate node (see Figure 3.1) has just been activated. Being most primed (usually), this primed-from-above node reaches threshold sooner than other "extraneous" nodes in its domain and becomes activated.

Once a content node becomes activated, its sequence node must return quickly to resting level, because content nodes have a return connection to their sequence node, which could cause reverberatory reactivation. Thus, an activated content node must quench, or inhibit, rather than further prime, its corresponding sequence node, so that only one content node becomes activated at any one time. Without being quenched, a sequence node could simultaneously activate several nodes in its domain, causing a potential breakdown in behavior. Quenching requires a threshold mechanism, which if exceeded, causes content nodes to inhibit rather than prime their sequence nodes.

The Sequencing Function

It is important to understand that priming is fundamentally nonsequential. An activated content node primes all of its connected nodes at the same time. Sequence nodes are needed to impose the sequence of activation, and thereby determine the appropriate temporal sequence for words, segments, and muscle movements in the final output. For example, activating *practice*(noun, verb) simultaneously primes *prac*(stressed syllable) and *tice*(unstressed syllable), and as discussed above, the connections between sequence nodes represent which one comes first, enabling the correct sequence to be generated in the final output.

A Simplified Example

To illustrate how timing and sequence nodes interact to determine whether, when, and in what order content nodes become activated in everyday speech production, consider how the words *frequent* and *practice* might become sequenced in the noun phase "frequent practice." Figure 3.2 illustrates the

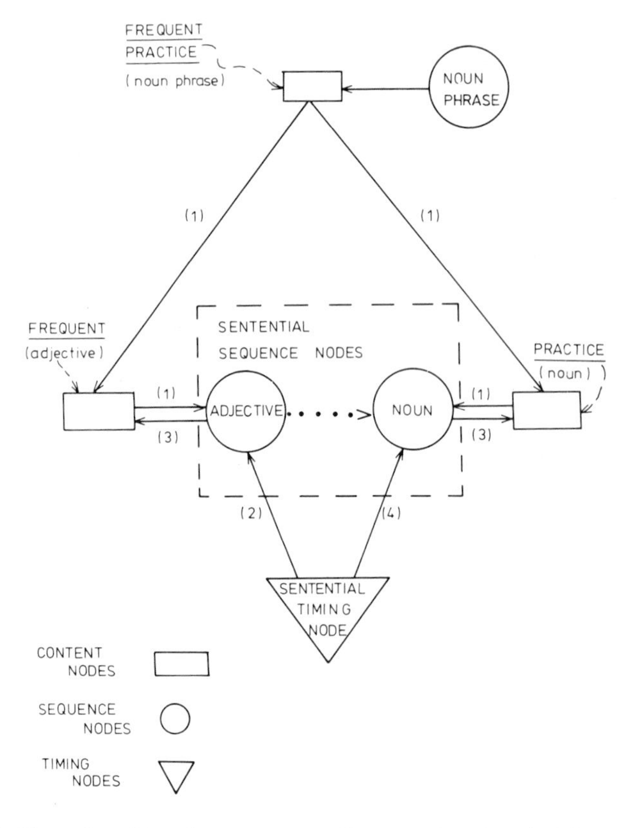

FIGURE 3.2. The order of top-down processes (in parentheses) underlying activation of content nodes (in rectangles), sequence nodes (in circles), and the sentential timing node (in triangle) for producing the noun phrase *frequent practice*. (From "The Problems of Flexibility, Fluency, and Speed–Accuracy Trade-Off in Skilled Behavior" by D. G. MacKay, 1982, *Psychological Review*, *89*, p. 492. Copyright 1982 by American Psychological Association, Inc. Reprinted by permission.)

hypothetical content nodes (in rectangles), sequence nodes (in circles), and timing node (triangle). Unbroken lines in Figure 3.2 are excitatory, the broken lines represent the quenching mechanism, and the dotted lines represent the inhibitory relationship between sequence nodes. Similar connections and processes are postulated for all sequentially organized mental nodes.

Assume that the node representing the sentential concept, *frequent practice*(noun phrase) has received top-down priming and is activated first in the same way as any other nodes, such as the ones described below. Activation of *frequent practice*(noun phrase) causes simultaneous priming of *frequent*(adjective) and *practice*(noun), which immediately pass on second-order priming to their hypothetical sequence nodes, ADJECTIVE and NOUN. The inhibitory link between ADJECTIVE and NOUN temporarily reduces the priming level for NOUN, so that ADJECTIVE is most primed and becomes activated under the most-primed-wins principle following the first pulse from the timing node. ADJECTIVE therefore multiplies the priming of every content node in its *adjective* domain, and the one with the most priming in the domain reaches threshold soonest and becomes activated (the most-primed-wins principle). The node with the most priming in the *adjective* domain will of course usually be *frequent*(adjective), which has just recently been primed by *frequent practice*(noun phrase).

Once activated, *frequent*(adjective) "quenches" its sequence node; that is, it quickly reduces the activity of ADJECTIVE to resting level, thereby ensuring that one, and only one, content node in the domain becomes activated at any given time. Quenching of ADJECTIVE releases the inhibition on NOUN, which therefore dominates in degree of priming within the domain of sentential sequence nodes, and NOUN becomes activated under the most-primed-wins principle following the next pulse from the sentence timing node. NOUN therefore multiplies the priming of the entire domain of *noun* nodes, but having just been primed, *practice*(noun) has more priming than any other node in the domain and becomes activated under the most-primed-wins principle.

This example gives a flavor of the complexity of the processes that must underly sequencing in a skilled behavior such as speech production. As already noted, however, I have chosen a simple example. It focuses on only two nodes among the millions of nodes that are relevant to the way words combine to form the grammatical sentences of English. And I have simplified this simple example even further by representing the domain of, say, the word *frequent*, as *adjective*. As discussed above, the domain of *frequent* must be *temporal adjective*, rather than just *adjective*.

Differences Between Sequence and Content Nodes

How do sequence and content nodes differ? Although sequence and content nodes are similar in some respects, they differ in others, and the differences summarized in the following section are important for understanding the theory.

Functional Differences

Two basic functions that differentiate sequence nodes from content nodes are activation and sequencing. Sequence nodes can cause their connected nodes to become activated but content nodes cannot. No matter how long a content node primes a connected node, activation cannot occur without the help of the triggering mechanism or sequence node specific to the primed node. With respect to the sequencing function, content nodes cause simultaneous, that is, cotemporal or nonsequential activity in their connected nodes. Priming from an activated content node is invariably transmitted to all of its connected nodes at the same time. Only sequence nodes give rise to sequential activity by activating content nodes in a predetermined order.

Quantitative Differences

Both content and sequence nodes are organized into domains, receive priming from connected nodes, and are activated by multiplication of priming under the most-primed-wins principle. However, there are *quantitative* differences in the connections and domains for sequence versus content nodes. A content node typically connects with a single sequence node and with only two or three content nodes, which usually occupy different domains. In contrast, a sequence node connects with up to a thousand content nodes, which always occupy the same domain. For example, NOUN only connects with content nodes in the domain *noun*. A sequence node also connects with other sequence nodes, but only ones in its own domain. For example, NOUN only connects with other sequence nodes in its own (sentential sequence) domain.

Content and sequence nodes also differ greatly in number. Because different sequence nodes by and large connect with different content nodes and because each sequence node connects with many (up to a thousand) content nodes, content nodes must outnumber sequence nodes by a ratio of up to a thousand to one.

The number of domains that content versus sequence nodes are organized into also differs by at least an order of magnitude. Content nodes are organized into as many domains as there are sequence nodes. In the case of speech production, there are hundreds of sequence nodes and, therefore, hundreds of content node domains. In contrast, sequence nodes for speech seem to be organized into only about three domains: sentential sequence nodes, phonological sequence nodes, and (speech) muscle movement sequence nodes.

In summary, sequence nodes can be considered semispecific rather than nonspecific activating mechanisms. Because sequence nodes connect with, and activate, a specific and limited set of content nodes in domains such as *color adjective*, they exhibit some degree of specificity. However, sequence nodes also exhibit considerable nonspecificity, and certainly more nonspecificity than content nodes, because they *always* connect with, and activate, a category of content nodes rather than just one.

Differences in Connections

Connections between one content node and another differ from connections between one sequence node and another. Connections from one content node to another are simple (nonmultiplicative) and excitatory in nature and usually involve nodes in different domains. For example, a noun phrase node usually connects with nodes in domains such as *adjective* and *noun*. In contrast, connections between sequence nodes are inhibitory in nature and always involve nodes within the same domain. For example, the sequence node ADJECTIVE inhibits NOUN within its own domain of sentential sequence nodes for producing English.

Connections from content to sequence nodes also contrast with connections from sequence nodes to content nodes. Sequence nodes only send a multiplicative connection to their connected content nodes, while content nodes send a simple excitatory connection and a quenching connection to their connected sequence node(s).

Centrality to a System

Sequence nodes can be considered more central to a system than timing and content nodes. For example, sequence nodes receive connections from all three types of nodes: timing nodes, content nodes, and other sequence nodes. Content nodes, although essential for all instances of behavior, are structurally more peripheral than sequence nodes and only receive connections from sequence nodes and other content nodes. Timing nodes are likewise essential, at least for skilled behavior, but are the least central type of node in a system. A timing node receives connections from neither sequence nor content nodes within its own system.

Implications of the Theory

The node structure theory was designed to incorporate the general requirements for theories of sequencing outlined in the introductory section of this chapter, and it should come as no surprise that the theory explains the phenomena discussed there. I therefore focus on only two of these phenomena, in order to illustrate some implications of specific applications of the theory.

Production Onset Time

Recall that production onset time (the time to begin to produce a preplanned behavior) is shorter when the behavior consists of a single component than when it consists of a sequence of components. The node structure theory explains this finding as due to the time required to prime and activate the nodes preceding the first muscle movement node in the sequence. Some of these activated nodes are sequence nodes, and onset time also depends on the interactions between sequence nodes, that is, the number of serial-order rules that must become activated (or, equivalently, sequential decisions that must be made; see Figure 2.5) prior to activating the first muscle movement nodes.

Temporal duration at the surface level is irrelevant to production onset time under the theory, so that, despite the large differences in surface duration, only small increases in onset time can be expected for a one- versus two-syllable word, for a one- versus two-sentence paragraph and for a one- versus two-topic preplanned lecture (all other factors except output duration being equal). Even the number of preplanned components per se is irrelevant to onset time. As already noted, production onset time for, say, *pain* and *paint* should be equivalent under the theory (all other factors except length being equal), because the same number of nodes and serial-order rules must become activated before activating the first muscle movement node for /p/. The *nt* node and the serial-order rule for n + t in *paint* only become activated after the first muscle movement node has been primed and activated. Here the extra nodes and sequential operations leave onset time unchanged and add only to the overall time to produce the word. This prediction contrasts sharply with predictions from other theories (e.g., Klapp, 1979), which view the temporal duration of a preplanned sequence as the critical determinant of production onset time.

The node structure theory also provides a coherent account of Klapp and Wyatt's (1976) onset time data for sequences of finger movement. Recall that production onset times are longer for Morse code sequences beginning with a *dah* than a *dit*. A *dit* requires three hierarchically organized content nodes above the muscle movement level (see Figure 3.3): the highest level "*dit node*," and two subordinate nodes, one representing the press, and the other the lift for releasing the key. In support of this representation, Wing (1978) showed that the mechanisms for pressing and lifting the finger from a key are independent. A careful analysis of the timing characteristics of the lift and press components in an experiment involving repetitive finger tapping indicated that lifting the finger from the key is not triggered by the preceding press or vice versa. As Wing (1978) points out, the independence of these components is also required to explain the blocks or temporary pauses that sometimes occur in rapid finger tapping and other repetitive activities (see also Glencross, 1974). These blockages suggest that the independent press and lift components fall progressively out of phase until they become activated simultaneously rather than sequentially at the muscle movement level, so that no movement can occur.

Note, however, that a *dah* is more complex than a *dit* by any analysis: a *dah* requires an additional node for representing the fact that the key must be held in contact with the terminal, plus a timing mechanism for specifying the duration of this contact phase. (Figure 3.3 leaves out the timing mechanism but represents one possible relation between the content nodes for a *dah*, although not the only possible relation given our current state of knowledge about Morse code.) The longer onset time for *dah* (in either first or second position) may therefore

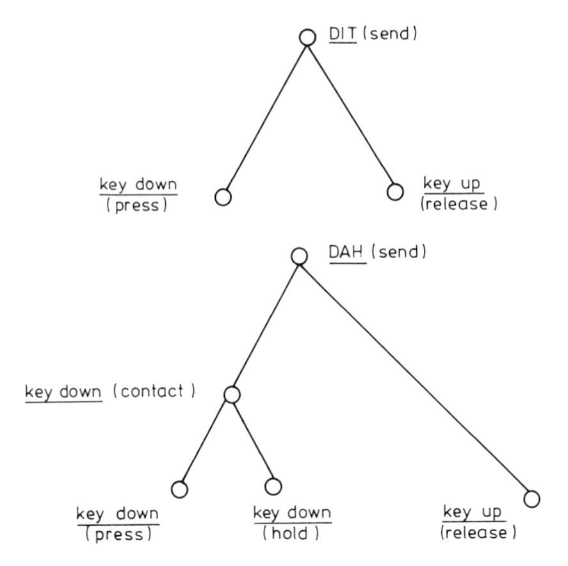

FIGURE 3.3. The structure of content nodes for a beginner producing *dit* versus *dah* in Morse code (D. G. MacKay, 1985).

reflect the greater number of content nodes, serial-order rules, and timing pulses that must be activated to begin producing a *dah*.

The theoretically more challenging finding concerns the longer initiation times for sequences containing different components (i.e., *dah-dit* and *dit-dah*) than for sequences containing identical components (i.e., *dah-dah* and *dit-dit*). Under the node structure theory, this finding reflects a difference in the mechanisms for repeating an element versus sequencing different elements. A repeated component requires a simple repeat mechanism, without any sequential decision (interaction between sequential nodes representing a sequential rule), whereas sequencing different components requires a (time-consuming) sequential decision involving interactions between two or more sequence nodes. Moreover, in this particular case, the two sequential rules conflict with one another (*dit* + *dah* for one sequence and *dah* + *dit* for the other) and cannot be called up in advance, adding further to the time required to resolve which element comes first under the most-primed-wins principle.

The Sequential Class Regularity

The sequential class regularity occurs in speech errors when one linguistic component inadvertently substitutes another. The substituted and substituting components almost invariably belong to the same sequential class. For example, over 99% of the word substitution errors in Stemberger's (1985) corpus obeyed the sequential class regularity. Nouns substituted with other nouns, verbs with verbs, and not with, say, nouns or adjectives. As noted earlier, this regularity also holds for other levels of speech production: for substituted morphological components (prefixes substitute with other prefixes, suffixes with other suffixes, and never prefixes with suffixes); for substituted syllabic components (initial consonant clusters substitute with other initial clusters, final with final, but never initial with final); and for substituted segmental components (vowels substitute with vowels, consonants with consonants, and never vowels with consonants).

The sequential class regularity also holds for other types of errors: anticipations, perservations, and transpositions (although somewhat less strongly; see the exceptions discussed in the following section). As Meringer and Mayer (1895) pointed out, a common mechanism almost certainly underlies all three classes of errors, and the most-primed-wins principle is exactly the sort of common underlying mechanism that is needed. Errors occur under the node structure theory whenever an "intended" node has less priming than some other "extraneous" node in its domain when the activating mechanism is applied. Because an activating mechanism also applies to a particular domain or sequential class, this means that substituted components at every level in the system will belong to the same sequential category. For example, if *crotch*(noun) acquires greater priming than *watch*(noun), for whatever (e.g., Freudian) reason, *crotch* will substitute for *watch* when the activating mechanism is applied to the noun domain. However, the noun *crotch* will never substitute with a verb, even a phonologically similar verb such as *botch*, because the corresponding activating mechanisms cannot be applied simultaneously to both the noun and the verb domains. Because of the most-primed-wins principle, only one sequence node can be applied at a time; either NOUN or VERB can become activated, but not both simultaneously. Of course, this is not to say that one initial consonant, /b/, cannot substitute for another, /w/, to cause the substitution of *botch* for *watch*. When this occurs, however, the error involves not a word substitution, but a phonological substitution which obeys a sequential class regularity of its own: initial consonants substitute with initial consonants.

The node structure theory predicts, further, that the sequential class phenomenon will hold statistically for *all* types of errors, including also nonsequential errors such as blends and malapropisms. The reason is of course that an activating mechanism can only activate and misactivate nodes within the same sequential domain as the appropriate or intended-to-be-activated node. A similar regularity should also hold for errors in other highly skilled behaviors, even those involving very different sequential classes, such as typing, where the two hands seem to make up one domain; the different types of strokes (horizontal, vertical, lateral) make up a second; and the homologous fingers of the two hands make up a third class of domains (Grudin, 1981).

Exceptions to the Sequential Class Rule

About 20% to 30% of Stemberger's 1985 corpus of sequential errors (anticipations, perseverations, and transpositions) failed to preserve sequential class, but these relatively infrequent exceptions to the sequential class rule often display regularities of their own that must be explained in theories of sequencing. Consider the following examples from Fromkin (1973), "She was waiting her husband for" (instead of, "waiting for her husband") and "I don't want to part this book with" (instead of, "to part with this book"). Errors such as these pose three questions: Why do they violate the sequential class rule? (Both errors involve a noun phrase changing places with a verb particle.) Why are they so rare? And why do they result in a sequence (verb + noun phrase + verb particle) that is appropriate for other expressions such as "She called the man up"? The node structure answer to these questions is that these errors reflect misapplication of a serial-order rule. Connections with the wrong sequence nodes (and therefore the wrong serial-order rules) have been formed, which results in the wrong order in the output. For example, a connection to *for*(particle) rather than, or even in addition to, a connection to *for*(preposition) could result in an error such as "waiting her husband for."

In short, the theory distinguishes two general classes of speech errors: *priming* errors and *rule* errors (see also Stemberger, 1985). Priming errors obey the sequential class phenomenon, and occur whenever an intended node has less priming than some other node in its domain at the time when the most-primed-wins activating mechanism is applied. Rule errors violate the sequential class phenomenon and occur whenever an inappropriate serial-order rule is called up or primed. Rule errors are relatively rare, because forming incorrect connections between nodes is relatively rare.

Of course the node structure theory doesn't yet provide a full or conclusive account of all aspects of speech errors. One of the many outstanding questions is why higher level nodes in a system are less prone to error than lower level nodes in the same system (Dell, 1985a). In the sentential system, for example, phrase nodes participate in fewer errors than do lexical nodes, and in the phonological system, syllable nodes participate in fewer errors than do segment or feature nodes. The answer is certainly *not* that higher level content nodes for phrases and syllables do not exist. After all, these units *sometimes* participate in errors. The answer suggested by the node structure theory is related to speed–accuracy trade-off. Nodes at higher levels in a system are less prone to error because they are activated at slower rates than lower level nodes in a system. For example, syllables may be less prone to error than segments or features because segments and features are produced much faster than syllables in the phonological system. A new syllable node is activated every, say, 500 ms, whereas a new segment node or set of feature nodes is activated every, say, 150 ms, allowing less time for priming to summate, and thereby increasing the probability of activating the wrong segment node under the most-primed-wins principle.

4
Perceptual Sequencing and Higher Level Activation

> The purpose of perception is not to produce an end-product (such as a percept), but to constrain actions in such a way as to continuously reveal useful aspects of the environment.
>
> (Michaels & Carello, 1981, p. 95).

> The much-worked claim that "illusions" and "failures of perception" are instances of failed inference . . . has about as much intellectual force as a cough in the night.
>
> (Turvey, Shaw, Reed, & Mace, 1980, p. 275).

Having examined the problem of sequencing in action, I turn now to the problem of perception, especially the problem of how we perceive input sequences in proper serial order when we do and improper order when we make errors. As in the previous chapter, I will begin with some general constraints that apply to any theory of perception and then construct a node structure theory of perception that incorporates these general constraints and makes predictions for future test.

General Constraints on Theories of Perception

Differences between sensation and perception, the problem of sequencing in perception, and basic phenomena such as perceptual constancies impose fundamental constraints on theories of perception. Needless to say, this chapter cannot comprehensively summarize all of the research related to these constraints. After all, perception has been the subject of many thousands of studies, and whole books have been written on categorical perception alone (Harnad, 1986), one aspect of the phenomenon of perceptual constancy. However, I do intend to touch on the basics and to put forward some strong claims as to their underlying basis.

Basic Phenomena

At least three basic phenomena must be explained in theories of perception: perceptual constancies, the category precedence effect, and effects of context on perception.

Perceptual Constancies

As Fodor (1983) points out, constancies of form, size, color, and phonology function as follows:

> . . . to engender perceptual similarity in the face of the variability of proximal stimulation. Proximal variation is very often misleading; the world is, in general, considerably more stable than are its projections onto the surfaces of transducers. Constancies correct for this, so that in general percepts correspond to distal layouts *better than* proximal stimuli do. . . . (Fodor, 1983, p. 60)

Constancies of phonology concern the fact that phonemes do not have an invariant acoustic representation in the speech signal. People hear different allophones or acoustic variants as the same phoneme. The output side exhibits an analogous problem. The actual movements associated with producing a phoneme vary with the contexts in which the phoneme is produced.

The mechanisms responsible for phonological constancy are also responsible for categorical perception, the fact that speech perception fails to follow a continuously varied stimulus but is categorical or discontinuous in nature. For example, when voice onset time for a stimulus resembling either a */d/* or a */t/* is varied along a continuum, the resulting stimuli are not perceived as continuously varying, but as belonging to one phoneme category or the other.

Quite diverse perceptual systems have been found to exhibit categorical perception: human music perception (plucked-string versus bowed-string violin notes; Cutting & Rosner, 1974); human color perception (Lane, 1965); and the recognition of speech stimuli by primates and chinchillas (Kuhl & Miller, 1975). Moreover, humans do not *necessarily* perceive phonemes categorically. Massaro and Cohen (1976; 1983a) showed that subjects can, if appropriately instructed, use acoustic features to discriminate between test stimuli that fall within a phonemic category (see also McClelland & Elman, 1986). Another interesting exception is the fact that normal length vowels do not exhibit categorical perception (e.g., Pisoni, 1975).

Phonological constancies also exhibit trading relations. Whenever two or more cues contribute directly to a phonological distinction, say, between a voiced stop versus an unvoiced stop, one cue can be traded against the other (within limits). If one cue in a synthesized syllable is changed to favor one alternative and the other cue is changed to favor the other alternative, the effects become integrated, and perception remains constant; the change in one dimension offsets the change in the other (e.g., Massaro, 1981; Massaro & Cohen, 1976; Summerfield & Haggard, 1977).

The Category Precedence Effect

The category precedence effect concerns the fact that subjects can sometimes perceive an entire category of objects or words (e.g., letters versus digits) faster than a particular member of the category. We can use the original experiment of Brand (1971) for purposes of illustration because more recent category search experiments (e.g., Prinz, 1985; Prinz, Meinecke, & Heilscher, 1985; Prinz & Nattkemper, 1985) corroborate the original results. Subjects in Brand (1971) were required to detect either (a) a particular digit embedded in a list of other digits or (b) *any* digit embedded within a list of letters. The results showed that response times were faster in condition (b) than in condition (a). *Any* digit among letters was identified faster than a particular digit among other digits.

How can perceiving that a character is a digit proceed faster than perceiving which particular digit it is? Is there some abstract and as yet unknown feature that distinguishes the *category* of letters from the *category* of digits? A follow-up experiment by Jonides and Gleitman (1976) ruled out this stimulus-based interpretation. The subjects were asked to detect either the *digit O* or the *letter O* as quickly as possible, with the *O* embedded either in a list of digits or in a list of letters. The *stimulus* for the letter *O* versus the digit *O* was therefore identical, but the results were as before. Subjects instructed to look for the *digit O* responded faster when the *O* was embedded within a list of letters than when it was within a list of digits. Conversely, subjects instructed to look for the *letter O* responded faster when the *O* was embedded within a list of digits than when it was within a list of letters. Category information (letter versus digit) can apparently facilitate perception independently of any possible surface feature for distinguishing one category from the other.

Context and the Part–Whole Paradox

As Fodor (1983) and others point out, the everyday fact that both prior and subsequent context facilitates the detection of letters, words, and objects is part of a theoretical paradox. Perception of a whole word, object, or scene seems to require perception of its parts, but at the same time is known to *influence* perception of its parts. Object contexts facilitate feature or line detection, and scene contexts facilitate object detection. For example, tachistoscopically presented objects are easier to identify when they form part of a real world scene than when they form part of a jumbled version of the same scene (Biederman, 1972). Similarly, a letter is easier to perceive within a familiar word than within an unfamiliar string of letters. The *D* in *WORD* is easier to perceive than the *D* in *WROD*, for example (e.g., McClelland & Rumelhart, 1981). Even acronyms induce a "word superiority effect." A letter is easier to perceive in a familiar letter string such as LSD or YMCA than in an unfamiliar letter string such as LSF or YPMC (Henderson, 1974). This finding cannot be explained in terms of bottom-up orthographic or phonological factors, and is paradoxical if identification of letters (parts) must precede identification of words (wholes). Resolving this part–whole paradox provides a basic challenge for theories of perception.

Differences Between Sensation and Perception

Differences between sensation and perception, such as those discussed in the following sections, impose additional constraints on theories of perception.

Sensations not Represented in Perception

As a general rule, at least some ongoing sensations are not represented in perception. We normally do not perceive the proximal stimulus, or full-blown pattern of sensory stimulation, but rather the distal stimulus or higher level conceptual aspects of an input. In speech perception, for example, we perceive and comprehend words but not low-level (e.g., phonemic and allophonic) characteristics of speech inputs. Similarly in visual perception, we perceive an object such as a lamp at some distance from ourselves, but we fail to perceive the disparity between the images in our two retinas that can provide the sole sensory basis for our distance judgment. Likewise in audition, we hear the sound of a car's horn as coherent and localized in space, but we fail to perceive the sensory events underlying this perception, for example, differences in arrival time of the sound to the two ears (e.g., Warren, 1982). Explaining why such sensations are not represented in perception, or more generally, why we perceive the distal, rather than proximal stimulus, imposes fundamental constraints on theories of perception.

Illusions: Perceptions not Represented in Sensation

As instances where perception fails to correspond to the input, illusions provide another challenge for theories of perception. The phonemic restoration phenomenon illustrates a typical illusion where an element missing in the input is nevertheless perceived. When an extraneous noise such as a cough or a pure tone is spliced into a magnetic recording so as to acoustically obliterate a speech sound in a word, the word sounds completely normal, and subjects are unable to tell which speech sound has been obliterated (Warren, 1970; 1982). For example, when subjects listen to a sentence containing the word *legi*∗*lature*, where a cough (∗) has been spliced in place of the /s/, the word sounds intact, and the missing /s/ sounds as real and as clear as the remaining acoustically present phonemes (Warren, 1970; 1982). The subjects somehow synthesize the missing /s/, and when informed that the cough replaced a single speech sound, they are unable to identify which sound is missing.

Effects of Unperceived Sensation on Behavior

A large number of findings indicate that sensations can influence behavior but at the same time fail to reach awareness. An example is the effect of allophonic variation on reaction time. Allophones are the set of perceptually indistinguishable acoustic variants of a phoneme. Subjects perceive all members of an allophone set as the same phoneme. Nevertheless, reaction time measures indicate that the acoustic differences between allophones undergo unconscious processing. When subjects make same–different judgments between acoustically presented pairs of phonemes, they are unaware of allophonic differences, but nevertheless respond "same" faster for two *identical* allophones than for two *acoustically different* allophones of the same phoneme (Pisoni & Tash, 1974).

As Fodor (1983) observed, findings of this sort are legion in studies of perceptual constancies, and these findings support the hypothesis that constancies of form, size, color, and phonology correct for the often misleading variability of proximal stimulation. Fodor also noted that "the work of the constancies would be undone unless the central systems which run behavior were required largely to ignore the representations which encode *uncorrected* proximal information" (1983, p. 60). What remains to be explained is how "the central systems which run behavior" ignore lower level variability when it comes to perception, but nevertheless respond to lower level variability when it comes to reaction time.

Sequencing in Perception

The problem of sequencing in perception is this: What mechanisms enable us to perceive and to represent input sequences in proper serial order when we do and in improper order when we make errors? Sequential perception presents as much of a challenge for psychological theories as does sequential behavior but has been relatively neglected. Studies of perception over the past century have concentrated mainly on static visual displays and have devoted relatively little attention to the perception of input sequences. In the following sections I discuss three general classes of phenomena that illustrate the problem of serial order in perception and impose constraints on theories of sequential perception.

Sequential Illusions

Sequential illusions occur whenever units coming later in an input sequence are perceived as coming sooner. Phonological fusions are one example. Phonological fusions occur, for example, when a subject wearing earphones is presented with an acoustic stimulus such as *lanket* in one ear followed by *banket* in the other ear. Even with a sizable (e.g., 200 ms) onset lag or temporal asynchrony between the stimuli, subjects often report hearing *blanket*, a fusion of the two inputs (Cutting & Day, 1975; Day, 1968). If perception accurately represented the input sequence, subjects would perceive the *l* followed by the *b*, because the input order at the acoustic level is *l* followed without overlap by *b*. Some subjects in fact do perceive the input sequence veridically, but there are large individual differences, and most subjects do not. Instead they fuse the inputs and report that the *b* preceded the *l* (Day, 1968).

As their name suggests, phonological fusions depend on a phonological rather than an acoustic representation of the input. Whereas phonological factors readily influence the probability of fusion, lower level factors within the acoustic analysis system do not. One of these phonological factors is "wordhood"; fusions are relatively rare when both inputs are words, but relatively common when both inputs are nonwords, such as *banket* and *lanket*. Words are also the most common type of fusion response, regardless of whether the stimuli are words or nonwords (Day, 1968).

Another phonological factor is sequential permissibility. Fusions always result in phonological sequences that are permissible or actually occurring within the listener's language. Percepts that violate phonological rules (e.g., *lbanket*) never occur, even when nonoccurring sequences represent the only possible fusions. For example, *bad* and *dad* never fuse, because nonoccurring sequences (*bdad* and *dbad*) are the only possible fusions.

By way of contrast, Cutting and Day (1975) found that acoustic factors have little or no effect on the likelihood of fusion. The probability of fusion remained constant when they changed the intensity and fundamental frequency of one of the fusion stimuli, or altered its allophonic characteristics by trilling an *r*.

The Perceptual Precedence of Higher Level Units

Theories of perception must explain why we sometimes perceive units that come later in an input sequence more quickly than units that come sooner. The time it takes to recognize segments versus syllables provides a good example. Subjects require less time to identify an entire syllable than its syllable-initial segment, even though the segment ends sooner than the syllable in the acoustic stimulus. The original experiment (Savin & Bever, 1970) can again be used for purposes of illustration, because many subsequent studies have replicated its basic findings and come to the same conclusion (Massaro, 1979). Savin and Bever (1970) had subjects listen to a sequence of nonsense syllables with the aim of detecting a target unit as quickly as possible. There were three types of targets: an entire syllable such as *splay*; the vowel within the syllable, that is, *ay*; and the initial consonant of the syllable, that is, *s*. The subjects were instructed to press a key as soon as they detected their target, and the surprising result was that reaction times were faster when the target was the entire syllable rather than either the initial consonant or the vowel in the syllable. Why do higher level units often take precedence in perceptual processing, and how in particular can a syllable or word be perceived before the phonemes making up the syllable or word?

Effects of Practice

Effects of practice represent a frequently overlooked constraint on theories of sequential perception. Warren and Warren (1970) noted that we can perceive the serial order of sounds in familiar words such as *sand* at rates of 20 ms per speech sound, but we require over 200 ms per sound for perceiving the order of unfamiliar sound sequences such as a hiss, a vowel, a buzz, and a tone (when recycled via a tape loop). One interpretation of these findings attributes this difference to practice or familiarity. Sequences of speech sounds are much more familiar than sequences of nonspeech sounds such as *hiss-vowel-buzz-tone*. Another interpretation focuses on acoustic differences between speech versus nonspeech sequences (Bregman & Campbell, 1971). However, this second interpretation will not do for Warren's (1974) demonstration of how practice facilitates the recognition of nonspeech sequences. Subjects in Warren (1974) repeatedly listened to initially unrecognizable sequences of nonspeech sounds such as *hiss-vowel-buzz-tone*, and after about 800 trials of practice, the subjects became able to identify the order of these sounds with durations of less than 20 ms per sound. Theories of sequential perception must explain this order-of-magnitude effect of practice on sequence perception.

The Node Structure Theory of Perception

In my development of the theory so far, I argued that some of the nodes for perception and action are identical, and I illustrated some of the interconnections between these mental nodes for perceiving and producing speech. I then examined how mental nodes become activated in proper sequence during production. I turn now to the issue of perception: Which priming and activation processes involving mental nodes give rise to perception, and can the node structure theory handle the constraints on theories of perception previously discussed?

Priming is necessary for activation, and activation is necessary for perception, and I begin by discussing how activation takes place in perception. I then examine a general principle in the node structure theory whereby many of the nodes in a perceptual hierarchy only become primed rather than activated, and therefore never give rise to perceptual awareness. Various sources of evidence for this "principle of higher level activation" are discussed. Finally, I apply the node structure theory to the constraints imposed by the problem of serial order in perception.

The Most-Primed-Wins Principle in Perception

As noted in Chapter 1, the dynamic properties and activating mechanisms for mental nodes are identical in perception and production. During perception, activation within a system is sequential, requires a special activation mechanism (sequence node), and takes place at a rate specified by a timing node. However, the main sources of priming arrive bottom-up during perception, rather than top-down, as during production. Consider, for example, how the node *frequent*(adjective) becomes activated following presentation of the word *frequent* in perception. Sensory analysis and phonological nodes converge (many-to-one) to provide strong bottom-up priming to their connected nodes, and this priming summates on *frequent*(adjective), which then transmits second-order priming to ADJECTIVE, just as in production. With the next pulse from the timing node, this second-order priming enables ADJECTIVE to become activated as the most primed sequence node. Once activated, ADJECTIVE then multiplies the priming of all nodes in the *adjective* domain, but only the most primed one, normally *frequent*(adjective) in the present example, reaches threshold and becomes activated.

The mechanisms underlying the most-primed-wins principle therefore apply in the same way to activate nodes in both perception and production. This most-primed-wins principle is especially important for explaining temporal context effects, where perceiving an ongoing input both influences and is influenced by what comes in earlier and/or later (e.g., McClelland & Elman, 1986; Salasoo & Pisoni, 1985; Warren & Sherman, 1974). When an activating mechanism is applied to some domain in the system, the most primed node always becomes activated, regardless of whether its priming arrived before, during, or after the current surface input; but because the activating mechanism for perceiving a unit is normally applied long after the unit has come and gone in the surface input (see the following discussion), both left-to-right and right-to-left context effects are to be expected under the theory.

In summary, the record or trace of an input in the node structure theory is duplex in nature (rather than unitary but malleable as in McClelland & Elman, 1986). There are two records, a priming record and an activation record. The activation record is all-or-none and self-sustaining (for a set period of time), has relatively permanent effects, and gives rise to perception. The priming record is graded, malleable, and temporary, does not self-sustain, and does not necessarily give rise to perceptual awareness. Priming decays over time when the activity of its connected (e.g., contextual) sources stops; it summates over time as long as its connected (e.g., contextual) sources of input remain active; and it becomes erased following activation of a node by its activating mechanism.

Activation of a node of course leaves intact the (rapidly decaying) priming record of the large number of other nodes that happened to have less than most-primed status at the time when the activating mechanism was applied. As we will see in Chapter 7 (and in D. G. MacKay, 1987), the priming–activation distinction provides a natural account of "subliminal effects" in perception, such as those seen in studies of ambiguity. This "dual-trace" aspect of the node structure theory contrasts sharply with McClelland and Elman's (1986) TRACE theory, where a single, unfolding record gets "crunched" (destructively altered) at regular intervals, say every 25 ms, on the basis of available "right and left" context. Interestingly, Lashley (1951) pointed to the source of evidence that may eventually distinguish between these two accounts: garden path sentences such as "Rapid writing/righting with his uninjured hand saved from loss the contents of the capsized canoe" (discussed in Chapter 2). The node structure theory predicts that the perceptual switch from "writing" to "righting," which occurs at the end of this sentence, will be very rapid, because by then the node representing "righting" will have switched from less-than-most-primed to most-primed status, and can be activated immediately. Under other theories, however, the switch from "writing" to "righting" requires a time-consuming reanalysis of the entire sentence.

Perceptual Invariance and Categorical Perception

Why does phoneme perception tend to remain invariant across a variety of acoustic signals, so that we normally hear different allophones or acoustic variants of a phoneme as identical? The reason is that segment nodes receive bottom-up connections from a large set of acoustic analysis nodes, subsets of which characterize different allophones, or context-dependent acoustic variants of the segment (Figure 4.1).

Now, one allophone can be considered prototypical (see also Massaro, 1981) and provides a better (e.g., less error-prone) acoustic stimulus for perceiving the phoneme, because it contributes more bottom-up priming than any other allophonic variant. However, differences between allophones in the bottom-up priming of segment nodes are normally never perceived in the theory because perception requires all-or-none activation. Under normal (error-free) conditions, the same segment node will invariably become most primed in its domain and activated (perceived), regardless of which of its allophonic variants is present in the acoustic input.

As we will see, the acoustic and phonological feature information underlying phoneme identification normally becomes primed, but not activated (the principle of higher level activation discussed later in this chapter), and certainly never (consciously) perceived. Most other theories also characterize the feature information underlying phoneme identification as inaccessible (see the motor theory of Liberman et al., 1962) and/or rapidly lost (see the dual code theories of Massaro, 1981; Pisoni, 1975). What the node structure theory does is provide a much more general basis for both the rapid loss (decay of priming) and the inaccessibility (nonactivation under the principle of higher level activation, described later) of feature information that these other theories simply assume.

The same principles of the theory explain an analogous phenomenon on the output side: the fact that movements associated with producing a phoneme vary with the contexts in which the phoneme is produced. Although a single node represents a given segment in both perception and production, any given segment node is connected to many different muscle movement nodes representing acoustic variants of the phoneme. Context-dependent priming arising within the muscle movement system then determines which of these muscle movement

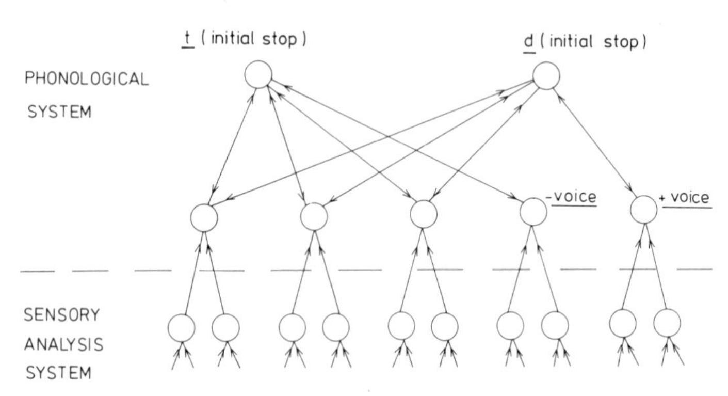

FIGURE 4.1. Connections from sensory analysis and phonological feature nodes to two segment nodes, representing syllable-initial /*t*/ and /*d*/.

muscle movement system then determines which of these muscle movement nodes becomes activated (D. G. MacKay, 1982), thereby introducing context-dependent motoric variation.

The same principles also explain why perception of speech sounds is generally categorical or discontinuous in nature. Stimuli varied along a perceptual continuum, such as voice onset time, are perceived discontinuously as belonging to one of two categories, such as /*b*/ or /*p*/, because at some point along the continuum, most-primed status will suddenly switch from one node to another within the relevant domain. For example, if an acoustic input resembles either a /*b*/ or a /*p*/, the /*b*/ node will receive more priming than the /*p*/ node when voice onset time is short, giving rise to perception of /*b*/ when the most-primed-wins principle is applied to the *stop consonant* domain. But when voice onset time is lengthened to the point where the /*p*/ node receives more priming than the /*b*/ node, a sudden discontinuity or categorical shift to perception of /*p*/ will occur. (See also McClelland & Elman's 1986 account, which is similar in some respects and goes into other aspects of categorical perception.)

Because most-primed-wins is a universal activating principle, applying at every level in every system, the phenomena of categorical perception and perceptual invariance should be universal as well. Although I have mainly used examples from speech perception to illustrate these phenomena here, the node structure theory predicts similar phenomena in other areas of perception, such as vision, touch, and music perception. The fact that categorical perception can be shown for human color perception, or for any other perceptual modality, is compatible with the theory. So is the fact that primates and chinchillas can perceive speech stimuli categorically, although it seems likely that their categories are acoustic rather than phonetic or phonological, and it remains to be explained why their acoustic nodes exhibit categorical sensitivity to voice onset times characteristic of English consonants.

Of course, the fact that synthesized vowels do *not* show categorical perception at first sight seems embarrassing for a general principle such as most primed wins. However, vowels, unlike consonants, can be described as musical chords, and if subjects are treating the vowels in categorical perception experiments as music rather than as speech (i.e., activating nodes representing patterns of pure tones), it is not surprising that these subjects fail to show boundary effects corresponding to English vowels; the categorical boundaries for acoustical tones are much narrower than those for speech. This analysis predicts categorical effects within a much narrower frequency range for subjects instructed to treat vowels as music, a prediction not shared by other theories (e.g., Pisoni, 1975) that assume inherently more persistent priming or short-term memory for vowel features than for consonant features.

Trading Relations

Under the node structure theory, trading relations illustrate how priming from lower level nodes summates at higher level nodes. An increase in bottom-up priming arriving from one set of nodes that is sufficient to offset a decrease in priming arriving from another set of nodes will leave summated priming unchanged, so that the same segment node will be activated under the most-primed-wins principle, and perception will remain the same. Thus, if one cue in a synthesized syllable is changed to favor a voiced stop, and the other cue is changed to favor the corresponding unvoiced stop, summated priming remains the same as if no changes had been made; the change in one dimension offsets the change in the other. Needless to say, there are many other accounts of the trading-relations phenomenon (McClelland & Elman, 1986), and some have provided more detailed fits to the empirical data (e.g., Massaro, 1981). However, the node structure account differs from other accounts in several fundamental respects, such as the distinction between priming versus activation, and these differences must eventually be subjected to empirical test.

The Specialness of Speech

Like the motor theory of speech perception (Liberman et al., 1962), the node structure theory recognizes the specialness of speech among systems for perception and action. Speech systems are activated independently from other perception and action systems in the node structure theory. For example, one and the same auditory stimulus can be analyzed as a speech event, by activating the sequence nodes for the phonological system, or as a nonspeech event, by activating the sequence nodes for the auditory concept system. The staggering degree of practice that speech normally receives (D. G. MacKay, 1981, 1982) also makes speech special in the node structure theory, as does the self-inhibitory mechanism that content nodes for speech require to deal with self-produced feedback (Chapter 8). Whereas speech stimuli can be self-produced, people cannot self-produce visual stimuli such as the external world—except marginally in drawing, typing, writing or moving the eyes—and this means that not all systems representing the visual world require self-inhibitory mechanisms.

However, speech is not *fundamentally special* in the node structure theory. Similar node structures and degrees of practice can be achieved in principle, if not in practice, within other perceptual and motor systems. Moreover, although different speech and nonspeech systems and modalities differ in nodes, and perhaps also node structures or patterns of connections (D. G. MacKay, 1987), they do not differ in fundamental principles of activation.

The Category Precedence Effect

Why are categories of words and objects sometimes perceived faster than particular members of the category? In Jonides and Gleitman (1976), for example, how did subjects identify *any* digit among letters faster than a particular digit among other digits? This finding is paradoxical if it is assumed that identifying a particular exemplar is necessary for identifying its category. Dick (1971) deepened the paradox by showing that subjects can name a visually presented character from

100 ms to 200 ms faster than they can discriminate that the character is a digit versus a letter. Deepening the paradox further, Nickerson (1973) showed that identifying a character as a digit versus a letter and identifying what the character is required the same quality of information. For stimuli presented in noise, subjects failed to distinguish between a digit versus a letter unless they could also decipher which digit or letter it was (see also Prinz & Nattkemper, 1985).

Can the node structure theory explain all these seemingly contradictory findings? The basic category precedence effect follows directly from the speed–accuracy trade-off postulate of the theory (Chapter 2; and D. G. MacKay, 1982). Subjects activate the most primed node in the digit domain when instructed to look for a digit and the most primed node in the letter domain when instructed to look for a letter. Thus, with the *O* embedded in digits, errors are more likely when subjects are instructed to look for the *digit O* than for the *letter O*, because activating the most primed node in the letter domain will not suffer interference (an increase in the probability of errors) when irrelevant (extraneous) nodes in the number domain become primed, but activating the most primed node in the number domain *will* suffer interference. Because speed trades off with errors, this means that, with errors held constant, detecting the digit *O* among letters will take longer than detecting the letter *O* among letters.

However, the quality of information required to detect the *O* versus to classify the *O* as a letter or as a digit will be identical in the theory. Priming transmitted from a content node to its sequence node provides the basis for classification, and also provides the basis for identification, which can only occur when the target content node has greater priming than any other node in its domain. Moreover, the process of naming a letter is different and more direct than the process of generating a proposition such as "*O* is a letter," and it is not at all surprising that the naming process is faster.

More generally, between-category searches in the node structure theory can make use of existing or preformed connections between sequence and content nodes. To successfully detect a digit among letters, for example, the activating mechanism for the domain of digit nodes can simply be applied over and over on each trial. This general strategy has two consequences. One is that a node in the digit domain will become activated soon after it becomes primed, enabling the already discussed rapid recognition response. The other consequence is a high probability of "false alarms" to nontargets in the same category as the target. Repeated application of a sequence node will automatically activate whatever content node has greatest priming in the domain or category, regardless of whether the priming arises from a target or a nontarget. This explains an interesting finding of Gleitman and Jonides (1976) that subjects searching for a particular digit among letters respond incorrectly with very high probability to "catch trials" with a nontarget digit. For example, subjects instructed to search for a *3* among letters often respond "present" to a catch trial digit such as *6*. The reason is that the content node representing *6* will automatically become activated as the most primed digit node if the sequence node for digits is repeatedly applied.

Unlike between-category search tasks, within-category search tasks in the node structure theory cannot make use of existing connections. Successful detection of a *particular* digit embedded among other digits requires formation of new connections. A new domain, consisting in this case of a single node representing the target digit, must become established with its own special activating mechanism. Considerable practice is needed to strengthen the connections between sequence and content nodes within this new domain so as to achieve rapid reaction times for detecting the target digit among other digits. Once this practice has taken place, however, the node structure theory predicts disappearance of the category precedence effect. That is, with extensive practice, a particular digit among other digits can become as easy to detect as the same digit among letters. A similar process is required for explaining why consistent mapping conditions are superior to varied mapping conditions in the visual search tasks of Shiffrin and Schneider (1977).

The Principle of Higher Level Activation

Activation is *necessary* for perceptual awareness in the node structure theory but not *sufficient*. An additional mechanism, discussed in detail elsewhere (D. G. MacKay, 1987), is required in the theory for achieving perceptual awareness. This "consciousness mechanism" only complicates the present discussion, however, and in order to simplify exposition, I will pretend that activation is synonymous with perceptual awareness in the pages to follow.

As noted in Chapter 2, not all nodes in a bottom-up hierarchy, such as the one in Figure 2.4, become activated during perception, the way they do in a top-down hierarchy during production. Only higher level nodes normally become activated and give rise to everyday perception. In particular, I will argue that only nodes above the phonological system become activated in perceiving everyday speech. This "principle of higher level activation" is extremely general and will be illustrated with examples from visual and auditory perception, as well as from speech perception.

I begin with the logical basis for the principle of higher level activation, the fact that activating lower level nodes is *unnecessary* in perception. I then show why activating lower level nodes is *undesirable*, and I discuss the optimal level for activation to begin during everyday speech perception. Finally, I discuss various phenomena whose explanation seems to require a principle of higher level activation.

Why Lower Level Activation is Necessary in Production

To understand why activation at lower levels is unnecessary in perception, it helps to examine the reasons why activation is necessary *at all levels* in *production*. Two reasons stand out. One is that even lower level components such as segments must be produced in sequence, so that nodes *must* become activated in production, rather than just primed, because activation is sequential whereas priming is not. The corollary fact that segment units activated during production never reach awareness (except when an error occurs; see Chapter 9) further illustrates the need for a theoretical distinction between activation and awareness.

The commitment threshold is the second reason for activating nodes at all levels in production. Recall from Chapter 1 that a minimum level of priming, designated the commitment threshold, is required to activate a node. To become activated, content nodes must not only be most primed in their domain; their priming must reach commitment level, so that multiplication of priming via the sequence node can reach activation threshold and activate the node.

In production, then, higher level nodes must pass on sufficient priming to reach the commitment threshold of the lowest level nodes in an action hierarchy in order for behavior to occur. However, because top-down connections are one-to-many, that is, they diverge rather than converge, priming from highest to lowest level nodes cannot summate during production. This means that, without activation along the way, priming transmitted to the lowest level muscle movement nodes would fall below commitment threshold.

Why Lower Level Activation is Unnecessary in Perception

Transmission of priming is quite different in perception than in production. For temporal and structural reasons discussed in the following paragraphs, lower level nodes, without themselves becoming activated, *can* pass on sufficient bottom-up priming during perception to enable higher level nodes to accumulate enough priming to reach commitment threshold and (indirectly) become activated.

Why is *bottom-up* priming passed on so efficiently? Three fundamental factors play a role. One is the fact that bottom-up connections are convergent or many-to-one (Figure 4.2). Because convergent connections allow priming to summate, lower level nodes, without themselves becoming activated, can combine or converge to pass on sufficient bottom-up priming to reach the commitment threshold of their connected nodes.

Linkage strength also contributes to the efficiency of lower level bottom-up priming during perception. Lower level connections have greater linkage strength than higher level connections (D. G. MacKay, 1982), so that priming can be efficiently transmitted via these strong bottom-up connections between lower level nodes, without the help of activation along the way.

Temporal parameters are the third factor contributing to the efficient summation of lower level bottom-up priming. Convergent priming from lower level connections is simultaneous, or nearly simultaneous. Two or more lower level nodes prime a connected node at identical or nearly identical times during speech perception. For example, all four feature nodes illustrated in Figure 4.2 prime their connected segment node at the same time. In contrast, higher level nodes generally prime a connected node at different and nonoverlapping times.

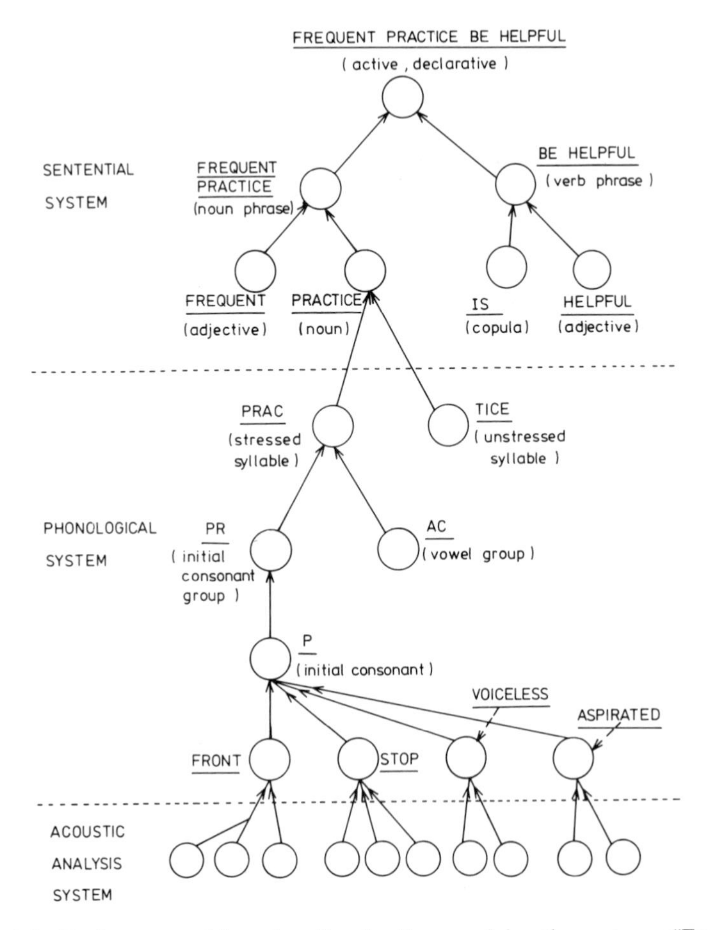

FIGURE 4.2. The bottom-up hierarchy of nodes for perceiving the sentence "Frequent practice is helpful."

Consider the arrival of convergent priming to the proposition node in Figure 4.2, for example. Priming from the verb phrase node will normally begin about 2 or 3 s after priming from the noun phrase node, and by that time, priming from the noun phrase node will already have begun to decay. In general, time lags between priming inputs will increase with the level of a node in the hierarchy, so that decay of priming will present more of a problem for higher than lower level nodes. As a result, lower level nodes generally receive greater temporal summation of priming than higher level nodes.

To summarize, higher level nodes must become activated during perception in order to transmit enough priming for connected nodes to reach commitment threshold and, indirectly, to become activated under the most-primed-wins principle. However, even when they do not become activated, lower level nodes pass on enough (second-order) bottom-up priming for their connected (higher level) nodes to reach commitment threshold and, indirectly, to become activated.

Activating these lower level nodes is therefore unnecessary, and this constitutes a preliminary or logical basis for the principle of higher level activation.

To illustrate this logical basis in greater detail, consider the bottom-up connections to the word *practice* in Figure 4.2. To facilitate exposition, assume that the sensory analysis nodes representing the acoustic input provide the equivalent of first-order priming to phonological feature nodes. Without becoming activated, each feature node therefore passes on somewhat weaker (second-order) priming to its connected segment nodes. However, because each segment node receives bottom-up connections from at least four feature nodes, the second-order priming from all four feature nodes may summate to at least the level of first-order priming from a single node. The segment nodes transmit this summated priming to their connected phonological compound and syllable nodes, and again, because of convergent summation, temporal overlap, and high linkage strength, the combined degree of second-order priming may remain comparable to that of first-order priming from a single *activated* node. Because first-order priming invariably suffices to meet the commitment threshold and, in fact, constitutes the normal basis for activation during production, activating lower level nodes is unnecessary in perception for transmitting sufficient priming to higher level nodes.

The efficient transmission of lower level bottom-up priming is only one reason why activating lower level nodes is unnecessary in perception. Another reason is that sequential perception is unnecessary for highly practiced, lower level sequences of components. Consider perception of the phonemes in the word *legislature*, for example. As long as the higher level node, *legislature*(noun), can become activated, it is always possible to reconstruct (top-down) what the lower level sequence of phonemes "must have been." I will illustrate the details of this reconstruction process later in the chapter when I discuss the phonemic restoration phenomenon, where a phoneme can be absent from an input sequence, but nevertheless perceived in sequence by top-down reconstruction, that is, by priming resulting from activation of its lexical content node.

Why Unnecessary Activation is Undesirable

Up to now, I have been arguing that activation of lower level nodes is *unnecessary* in perception: priming suffices. I now carry the argument a step further by noting that unnecessary activation is undesirable and should not occur. The main reason is that activation is more time consuming than priming and should be avoided, if possible, in order to speed up the rate of perceptual processing.

Why is activation so much slower than priming? Two reasons stand out. One is the temporal bottleneck caused by the self-inhibitory process that follows activation but not priming (Chapter 8). The other is the sequential rather than parallel nature of activation. An activating mechanism (sequence node) must first receive a buildup of priming and then become activated via a pulse from its timing node. Only then can it activate its most primed content node via multiplication of priming. Activating more than one node at a time is virtually impossible, and rate of activation must not be so fast as to induce errors in either perception or production (D. G. MacKay, 1982). This further reduces the rate of activation, relative to priming. Activation may also require more energy or effort than priming, and this may make unnecessary activation even more undesirable.

Errors are another reason for not activating lower level nodes in perception. Perceptual inputs are much more ambiguous at lower levels than at higher levels, where ambiguity is defined in terms of the relative degree of priming of "intended" versus extraneous nodes within the same domain (see also Chapter 7). So defined, ambiguity is a major cause of errors at all levels of perception. For example, the phonological feature nodes representing + versus − consonantal will both receive some degree of priming at the point when the /s/ in the word *legislature* occurs, because acoustic cues for consonants and vowels overlap in the acoustic signal (see McClelland & Elman, 1986, among others). Input at the phonological feature level can therefore be considered relatively ambiguous, and activation of phonological feature nodes could easily result in error, that is, activation of the wrong feature node, − consonantal in this example. The resulting misperception and awareness of error would disrupt perceptual processing, making activation of phonological nodes undesirable. However, the probability of error drops sharply at higher (e.g., lexical) levels, because unlike acoustic cues for phonological features, cues for different words rarely overlap (McClelland & Elman, 1986). Higher level activation also contributes "noise resistance." An extraneous sound could completely mask the /s/ in *legislature*, for example, without changing the most primed status of *legislature*(noun), because no other node in the *noun* domain could receive comparable bottom-up priming and become activated in error (see following discussion).

Another reason for not activating lower level nodes in perception concerns one of the most fundamental purposes of perceiving: to generate adaptive action. I argue below that perception (i.e., activation) of low-level components can interfere with, rather than promote, adaptive action. I begin with the observation that actions based entirely on low-level components are neither necessary nor desirable in everyday human behavior. For example, consider the phonological nodes for producing and perceiving speech in a normal, turn-taking conversation. Activating phonological nodes during perception primes muscle movement nodes, in effect preparing the muscle movement system for *producing* the just-perceived sequence of phonemes. However, immediately repeating a just-heard phonological sequence is by and large neither necessary nor desirable in everyday conversations. What is normally required is a new and adaptive response, rather than a repetitive one, and activing phonological nodes representing the previous input could only slow down or interfere with such a response. On the other hand, activating higher level (e.g., lexical content) nodes provides the primary basis for forming new connections (D. G. MacKay, 1987), not just within the sentential system, but within other systems representing visual cognition, for example, and generating adaptive rather than repetitive responses generally requires the formation of new connections.

The Optimal Level for Activation

To summarize, activation incurs costs and benefits. Although activation costs time, and perhaps also effort, it enables perceptual awareness, which is desirable at the highest possible levels to ensure adaptive action. It follows from this analysis that activation will become cost-effective at some optimal level in the hierarchy. Below the optimal level, costs of activation (time, errors, and effort) outweigh benefits, and above the optimal level, benefits of activation (perceptual awareness and adaptive action) outweigh costs.

What is the optimal level? At what level should activation begin during everyday speech perception for example? I will argue that lexical content nodes represent this optimal level, at least for adults perceiving familiar words under favorable acoustic conditions. Consider first the degree of priming arriving at lexical content nodes, relative to higher level phrase and proposition nodes. Bottom-up linkage strength, temporal summation, and convergent summation from phonological nodes is so great that second-order priming alone can be considered sufficient to meet the commitment threshold of a lexical content node. For example, most lexical nodes have undergone thousands, and sometimes many millions, of prior activations over the course of a lifetime (D. G. MacKay, 1982), so that the strong bottom-up connections to these lexical nodes will transmit sufficient second-order priming to reach commitment threshold and permit activation.

In contrast, phrase nodes normally receive insufficient bottom-up priming unless their connected lexical nodes become activated and contribute additional, first-order priming. Linkage strength of bottom-up connections to phrase nodes is relatively weak, because phrases, like propositions, are by and large new and receive much less practice than lexical units. Most phrase nodes have undergone very few prior activations, and many have undergone none whatsoever (D. G. MacKay, 1982). As a result, second-order priming will normally fall below commitment threshold of phrase nodes. Activating lexical content nodes therefore becomes necessary for passing on sufficient priming to enable phrase nodes to become activated.

Another reason why lexical nodes are the first to require activation in perception is that words represent the first level where sequence cannot be stored in advance. As Chomsky (1957) pointed out, it is reasonable to suppose a memory representation for the sequence of phonological components making up a word, but it is unreasonable to suppose a similar representation for the sequence of words making up most sentences; there are just too many possible sentences to store them all. Thus, because activation is required for sequencing in the node structure theory, lexical content nodes must become activated in order to represent and, if necessary, retrieve the sequence of words in a sentence.

A final reason for activating lexical nodes first in perception is that lexical nodes are the first units in a bottom-up hierarchy that interconnect with mental nodes outside the language modality. In order to comprehend the word "apple," for example, *apple*(noun) must be connected to the visual concept nodes representing apples. Similarly, in order to name a visually perceived apple, visual concept nodes representing the apple must send a return connection to *apple*(noun). Thus, activating lexical nodes enables nodes in other mental systems to become strongly primed and activated so as to generate adaptive rather than "repetitive" responses to verbal inputs.

Flexibility of Higher Level Activation

Higher level activation is a relative rather than an absolute principle. As discussed in greater detail elsewhere (Chapter 5, and D. G. MacKay, 1987), there exists a mechanism for activating lower level systems of nodes, and this mechanism is called into play whenever an input is novel, degraded, or requires selective attention. These situations can be said to cause a downward shift in the cost-effective level for activation. Activating lower level nodes incurs costs such as reduced rates of processing, and perhaps also greater effort, but paying these costs is necessary in these situations in order to provide sufficient bottom-up priming to reach the commitment threshold of higher level nodes.

In any given experimental situation, some subjects may be more willing than others to pay the cost of lower level activation, and this may explain why studies using degraded or unfamiliar stimuli often exhibit large individual differences. These individual differences are sometimes the subject of unnecessary controversy. An example is the controversy over level of processing in studies of perceptual–motor adaptation (Repp, 1982). Some studies such as W. E. Cooper, Blumstein, and Nigro (1975, discussed in Chapter 2) obtained small but positive effects of perceptual–motor adaptation, and concluded that higher level units (mental nodes) common to perception and production were responsible for their results. Other studies failed to show perceptual–motor adaptation, and concluded that adaptation effects occur exclusively at an early stage in auditory processing, prior to phonological analysis.

Conflicting conclusions are also to be expected for less-than-optimal stimuli such as synthetically constructed nonsense syllables. If most of the subjects in one set of studies are analyzing the stimuli (in this case, activating nodes) at a sensory analysis level, whereas most of the subjects in another set of studies are analyzing the stimuli at the phonological level, conflicting results are inevitable. Individual differences in the level of activation may also be responsible for recent controversies over categorical perception (Massaro, 1981). Under the node structure theory, phonemes will be perceived categorically if phonological nodes alone become activated, but not if sensory analysis nodes also become activated. It is therefore not the case that subjects can *only* respond to speech stimuli in terms of absolute phonological categories. Subjects can apply the most-primed-wins principle *below* the segment level, even though they don't normally do this, and this unusual strategy enables perception of acoustic features for discriminating between test stimuli that fall within a phonological category, a phenomenon reported in Massaro and Cohen (1976) and elsewhere. Conflicting results associated with phonological fusions, discussed later in the chapter, also seem attributable to individual differences in level of activation.

Evidence for the Principle of Higher Level Activation

The principle of higher level activation does not apply during *production* in the node structure theory (unlike in Dell's, 1985b, theory). In order to *produce* the phonemes of a word in proper sequence, phonological nodes must invariably become activated. The principle of higher level activation is a *perceptual* principle, and several lines of empirical evidence can be shown to support the hypothesis that phonological nodes normally become primed but not activated during *perception*.

The Recognition of Segments Versus Syllables

The fact that it takes more time to identify segment targets than syllable targets (Savin & Bever, 1970) provides strong support for the principle of higher level activation. Such findings cannot be explained if all nodes in an input hierarchy must become activated or if activation of higher level nodes always requires activation of lower level nodes. Something like the principle of higher level activation is required. That is, the subjects initially must have activated only higher level (syllable) nodes, enabling rapid perceptual recognition of syllable targets. Perception of segment targets required an extra step, activation of segment nodes via multiplication of priming from the appropriate sequence node.

Reaction Times for Allophones

The principle of higher level activation explains the large body of findings indicating that lower level information can influence behavior (via priming), but nevertheless fail to reach awareness (which requires activation). An example is the effect of allophonic variation on reaction time. Even though all members of an allophone set are perceived as the same phoneme, unconsciously processed acoustic differences between allophones nevertheless influence same–different reaction times (Pisoni & Tash, 1974). Under the principle of higher level activation, only higher level (in this case, phonological, but not sensory analysis) nodes become activated, giving rise to perceptual awareness of phonemes, but not allophones. Some allophonic variants nevertheless prime their phoneme node more strongly than others, thereby enabling it to become activated more quickly (with error criterion held constant). However, sensory analysis nodes representing different allophones do not themselves become activated and give rise to perceptual awareness, so that allophonic variants influence reaction time, but not perception.

Perception of the Distal Stimulus

As expected under the principle of higher level activation, we normally perceive the distal stimulus, or higher level conceptual aspects of an input, and not the proximal stimulus, or pattern of sensory stimulation. In visual perception, for example, we perceive how far away an object is, but fail to perceive retinal disparities, the sufficient sensory basis for that perception. Similarly in audition, we hear the sound of a car's horn as coherent and localized in space, but we fail to perceive the sufficient sensory basis for localization, differences in arrival time of the sound at the two ears (Warren, 1982). The reason is that priming from the sensory analysis nodes representing sensory events is passed on so automatically, and so effectively, that full-fledged activation and perceptual awareness normally never occur at the sensory analysis level.

Noisy Input and the Phonemic Restoration Effect

Speech perception is remarkably insensitive to everyday noise and other input degradations (e.g., McClelland & Elman, 1986), and this efficiency is readily explained under the principle of higher level activation. For example, when an extraneous (nonspeech) noise such as a cough or pure tone acoustically obliterates a speech sound in a word, the word sounds completely normal, and subjects are unable to tell which speech sound has been obliterated (Warren, 1970). Subjects somehow synthesize the missing sound, and when informed that the cough has replaced a single speech sound, are unable to identify which sound is missing.

Phonemic restorations cannot be attributed to allophonic or coarticulatory cues in segments adjacent to the replacement sound because subjects restore the missing phoneme of a contextually appropriate word even when the extraneous sound has replaced an "incorrect" or deliberately mispronounced phoneme in a word. If allophonic cues were responsible for restorations, subjects should have perceived the *mispronounced* version instead of restoring the appropriate word (Warren, 1982). Moreover, the same missing segment can be perceived as many different phonemes, depending on the sentential context that precedes or follows the replacement sound (Warren & Sherman, 1974). Finally, similar restorations occur in other perceptual modules, where explanation in terms of coarticulatory cues is out of the question. For example, when an extraneous noise replaces a note in a familiar melody, the missing note undergoes perceptual restoration in the same way as the missing phoneme in a word (Warren, 1982).

Phonemic restorations are readily explained under the principle of higher level activation. For example, consider the sentence "The state governors met with their respective legi∗latures convening in the capital city" (from Warren, 1970). Lexical content nodes become activated first under the principle of higher level activation, and for the input *legi∗lature*, *legislature*(noun) will acquire greatest priming because of both bottom-up and top-down (right and left contextual) priming. Even though the cough (∗) has obliterated the *s* in the acoustic waveform, no other node in the (noun) domain is likely to acquire as much priming. *Legislature*(noun) therefore becomes activated under the most-primed-wins principle, and contributes top-down priming to its connected nodes, including, *is*(vowel group), and *s*(final consonant group). By applying the most-primed-wins principle to the (final consonant group) domain, *s*(final consonant group) can therefore become activated, causing clear perception of the obliterated *s*.

This is not to say that bottom-up priming arising from the replacement sound cannot influence the restorability of a speech sound. As Samuel (1981) points out, acoustic similarity between original and replacement sound plays a role in how readily the original sound is restored. With fricativelike white noise as the replacement sound, fricatives are better restored than vowels, but the opposite is true with a pure tone replacement. However, although acoustic similarity can influence restorability, it is not necessary for the occurrence of restoration under the node structure theory. Restorations should still occur for replacement sounds that are completely unlike any speech sound whatsoever.

Finally, I stress again that the principle of higher level activation only represents the *normal* processing strategy for perceiving and comprehending speech. Not all tasks elicit this normal strategy. In the task of phoneme monitoring, for example, subjects search for a particular speech sound in an incoming sentence and respond as quickly as possible after perceiving it. Here, activating the target phoneme on the basis of bottom-up priming represents a superior strategy to higher level activation, which would slow the subjects down. This observation is consistent with Foss and Blank (1980) and Foss and Gernsbacher's (1983) evidence that phoneme detection is basically a bottom-up process in the phoneme monitoring task. (See McClelland and Elman, 1986, for a related account that goes into greater detail on the issue of when top-down effects are and are not observed in speech perception.)

Context and the Part–Whole Paradox

Many findings can be used to illustrate the facilitative effects of context, including the just-discussed phonemic restoration phenomenon. Restorations of acoustically obliterated speech sounds occur when word and sentential context enables a lexical content node to become most primed and activated. Activation of the lexical content node in turn causes top-down priming of phonological nodes, enabling the node for the missing speech sound to receive greatest priming in its domain and become activated under the most-primed-wins principle. The result is clear perception of the missing speech sound replaced or acoustically obliterated by the cough.

More generally, two related aspects of the theory are required to explain facilitative effects of context on the detection of words and objects: (1) top-down priming arising from the identity of nodes for perception and production and (2) the most-primed-wins principle, which ensures that the most primed node becomes activated, regardless of whether it receives its priming from above or from below.

Consider now the part–whole paradox, which is the fact that perception of a whole word, scene, or object seems to require perception of its parts, but at the same time, can influence perception of its parts, as in the "word superiority" effect. Effects of the whole on perception of its parts are readily explained under the principle of higher level activation. Normally, only the higher level node representing the whole word or object becomes activated; lower level nodes representing parts only become primed. Moreover, perception of the whole only requires priming from *some* of its parts and never *requires* activation (i.e., perception) of any of its parts. Effects of the whole on subsequent perception of its parts are therefore unremarkable, because perception of the whole primes all of its parts top-down.

The part–whole paradox bears a theoretical relationship to the category precedence effect under the node structure theory. To illustrate this relationship, consider Bruner's (1957) demonstration that one and the same character can be perceived as a *B* among a sequence of letters but as a *13* among a sequence of numbers. How does context (numbers versus letters) bring about these differing perceptions? As in the category precedence phenomenon, such context effects are abstract in nature; the perceiver expects either numbers or letters *in general*, not a *specific* number or a *specific* letter. How can the abstract category of an unidentified stimulus precede and determine how the stimulus is perceived? Again the node structure theory provides a simple account of this and other examples of categorical context effects illustrated in Neisser (1976). The preceding (contextual) characters determine whether the activating mechanism (sequence node) for numbers or for letters becomes engaged, which in turn determines whether the most primed content node in the domain of the letters (B) or numbers (13) becomes activated, leading to perception of a letter versus a number.

Serial Order in Perception

What are the mechanisms whereby we perceive and represent input sequences in proper serial order when we do, and improper order when we make errors? The problem of perceptual sequencing has been virtually ignored in psychology, and is often considered trivial and uninteresting. The reason seems to lie in an implicit but fundamental assumption that has become built into virtually every theory of perception and memory published to date. Under this "*sequential isomorphism assumption*," perceptual sequences invariably mirror the external sequence of events in the real world: "first in" is "first perceived."

Sequential isomorphism usually holds in perception, but not always. Any theory of perception must explain why we usually perceive events in the order in which they occur, but there exist whole classes of striking and well-documented exceptions to this general rule, and we have already encountered several in the present chapter. I discuss the significance of these nonisomorphisms first, and then develop the node structure theory of sequential perception, show how it handles the nonisomorphisms, and examine some of its predictions for future test.

Violations of Sequential Isomorphism

Phonological fusions represent a clear violation of sequential isomorphisms. When presented with the acoustic stimulus *lanket* to one ear, followed 200 ms later with *banket* to the other ear, subjects should perceive the *l* followed by the *b*, given sequential isomorphism, because the order of arrival of the input is acoustic *l* followed without overlap by acoustic *b*. The fact that most subjects do not perceive the input this way, but instead fuse the inputs, and report that the *b* preceded the *l*, therefore violates sequential isomorphism (Day, 1968).

Phonemic restorations also violate sequential isomorphism. When subjects hear a sentence containing a speech sound masked by a cough (*), they are unable to accurately locate the cough within the sequence of phonemes, or tell which phoneme is missing when informed that the cough has physically replaced a single speech sound. The detection of clicks in sentences provides another paradigmatic violation of sequential isomorphism (see Fodor et al., 1974, for a general review). Finally, the fact that subjects can recognize syllables before syllable-initial segments (Savin & Bever, 1970) violates sequential isomorphism, because syllable onsets precede syllable offsets and so should be perceived first, given sequential isomorphism.

The Node Structure Theory of Sequential Perception

Perception of sequence depends on the sequence in which nodes become *activated* under the node structure theory and not on the sequence in which they become *primed*. Sequence in perception and in the external world can therefore exhibit nonisomorphisms. Although priming necessarily mirrors the detailed sequence of events in the real world, node activation does not, and node activation determines sequential perception. Because only higher level nodes normally become activated and give rise to perception (the principle of higher level activation), the priming of lower level nodes representing the sensory sequence doesn't necessarily determine the sequence perceived. For example, only the lexical content node for a familiar word normally becomes activated in perception, so that segments making up the word become primed but not activated and perceived in sequence.

Novel Sequences

Why is perception of rapidly presented and unfamiliar sequences so difficult? The main reason under the node structure theory is that perception of sequence requires activation. Priming is fundamentally simultaneous rather than sequential, at least for content nodes. A content node receives simultaneous priming from any number of other content nodes, with no indications as to sequence (D. G. MacKay, 1982). Reconstructing a sequence requires that sequence nodes become engaged, so that activation can occur, and this activation process takes time (see preceding discussion, and Chapter 7).

Why is the added time required for activation especially problematic for unfamiliar or novel sequences? One reason is that perceiving novel sequences requires activation of lower level nodes. The fact that a sequence is novel means that there are no nodes for representing the higher level components of the sequence. This means that sequential activation and perception must occur especially rapidly for novel sequences because inputs arrive more rapidly at lower levels than at higher levels. By way of illustration, compare the relative rates of activation required for word nodes versus segment nodes. If lexical nodes for words must be activated every 500 ms, on the average, phonological nodes for segments would have to be activated about five times as rapidly, say, one every 100 ms on the average. Without top-down help, then, perceiving a phoneme sequence requires activation at five times the rate that perceiving a word sequence requires. However, activation takes time and has a maximal rate, so that lower level sequential perception will break down at some rate of input where higher level sequential perception can still occur. Perceiving novel sequences therefore requires relatively slow rates of input because novel sequences require activation of lower level nodes (see preceding discussion).

As we will see, familiar sequences also enjoy the advantage of enabling listeners to reconstruct what the lower level sequence "must have been," and this reconstruction process is not possible for unfamiliar or novel sequences. Perceiving a novel sequence requires formation of an appropriate hierarchy of connections between content nodes, and each of these content nodes must become connected to a sequence node. Because forming these new connections requires additional time, and perhaps also extensive practice (D. G. MacKay, 1987), novel sequences can only be perceived at relatively slow rates of presentation or following extensive practice. The reason we perceive sequences of speech sounds so quickly and so effortlessly is that we have already had so much practice at doing so (D. G. MacKay, 1981; 1982).

Perceptual Lags

When the appropriate nodes and connections for representing a familiar input sequence such as a word have been formed and strengthened, the time course of perception is no longer locked into that of the input under the theory. The time to perceive becomes flexible, so that input and perception can proceed at different and variable rates, within limits. Indeed, because of the problem of ambiguity, discussed previously in this chapter and in Chapter 6, perception not only *can* but *should* lag behind the input by a considerable period.

How long a lag can be tolerated between input and perception? Limits to the lag are set by the degree of priming and its rate of decay for the nodes in question. Lags cannot be so long that priming decays below the commitment threshold of higher level nodes. The dichotic listening task illustrates the general nature of this limit. When subjects in dichotic listening experiments shadow one channel, or activate nodes representing what has been said on that channel, they can subsequently perceive what has been said simultaneously on the other channel up to several seconds earlier (D. G. MacKay, 1973a; 1987; Norman, 1969). This lag between input and perception is only possible because priming takes several seconds to decay. Sufficient priming remains after a few seconds so that the nodes representing information arriving on the other (unattended) channel can still be activated and give rise to perception. Of course, with delays longer than a few seconds, so much of the priming for an unactivated node will have decayed that activation and perception can no longer occur.

APPLICATIONS OF THE THEORY

I now reexamine the violations of sequential isomorphism, discussed earlier, in order to illustrate how the node structure theory handles the constraints these violations impose.

Phonological Fusions

How are phonological fusions explained under the node structure theory? Simultaneous presentation of forms such as *banket* and *lanket* automatically prime higher level nodes representing phonological compounds, syllables, and words, and using the principle of higher level activation (the normal strategy for everyday perception), the fusion, *blanket*, is the only possible perception. Because there are no lexical content nodes for *banket* or *lanket*, *blanket*(noun) will receive more priming than any other lexical node and automatically become activated. Even though *l* precedes *b* in the input, a fusion such as *lbanket* is impossible, because speakers of English don't have a node for representing, say, the syllable *lban*. A 200-ms temporal asynchrony between *lanket* and *banket* of course *can* be perceived at the sensory analysis level, but only by abandoning the principle of higher level activation and by adopting the unusual perceptual strategy of activating sensory analysis nodes.

The fact that appropriate sentential contexts increase the probability of fusing two simultaneous inputs is explained under the theory in the same way as other context effects. For example, the probability of fusing *pay* and *lay* to give perception of *play* increases in the context "The trumpeter will pay/lay for a while" (Cutting & Day, 1975). The reason is that the sentence context primes (top-down) the lexical node *play*(verb), which therefore acquires more priming than *pay*(verb) and *lay*(verb), so as to become activated under the most-primed-wins principle.

The node structure theory predicts that fusion responses will have greater frequency of prior occurrence than their input stimuli at the lexical, syllable, and phonological compound levels (all other factors being equal). The reason is that the lexical content node for a familiar word has connections with high linkage strength and accumulates more priming than the node for the potential fusion, which cannot therefore become activated under the most-primed-wins principle. This explains why common words tend not to fuse. Consider, for example, the dichotically presented stimuli *pin* and *sin* and their only possible fusion response, the lower frequency word, *spin*. When lexical content nodes for *pin*, *sin*, and *spin* become primed in the phonological fusion task, the high-frequency alternatives, *pin* and *sin*, because of the greater linkage strength of their connections, accumulate more priming than the lower frequency alternative, *spin*. As a result, the stimulus words, *pin* and *sin*, become perceived under the most-primed-wins principle, and not the potential fusion response, *spin*. Word stimuli fuse less often than nonword stimuli for a similar reason: Nonword stimuli, such as *lanket* and *banket* have no lexical content nodes that compete for priming with the fusion response (*blanket*).

Consider now the individual differences contributing to fusion versus nonfusion (Day, 1968). One hypothesis attributes these individual differences to the level at which nodes are becoming activated. Fusers are activating nodes at higher (lexical and phonological) levels, using the principle of higher level activation, whereas nonfusers are activating nodes at lower (sensory analysis) levels, that is, below the phonological level where fusions can take place. By activating nodes below the normal "level of processing" for everyday speech, nonfusers therefore achieve more accurate perception of the actual acoustic sequence.

Nonfusers could of course apply this strategy more generally to other input modalities, which may explain Keele and Lyon's (1982) demonstration that nonfusers for speech tend to be nonfusers for tones, accurately perceiving the order of tones presented simultaneously with a slight onset lag. Fusers for speech likewise tend to be fusers for tones; they experience difficulty determining which tone came first, perhaps because they are only activating higher level nodes in both systems. Under this "level of processing" explanation of these results, practice, feedback, and instructions such as, "pay attention to the sounds themselves" should suffice to transform these fusers into nonfusers.

An alternate hypothesis, suggested by Keele and Lyon (1982), holds that fusers have difficulty discriminating the order of onset for all stimuli, whether speech or nonspeech, because their timing nodes innately are less finely tuned. This being the case, the node structure theory predicts that fusers will display a timing deficit in both production and perception, because the same timing nodes control both (Chapter 5). Moreover, neither practice, feedback, nor instructions will eradicate the deficit.

Phonemic Restorations

I have already discussed the node structure account of why a phoneme that has been masked by an extraneous sound such as a cough (∗) sounds as real and as clear as the remaining acoustically present phonemes, even when subjects have been informed that the cough has physically replaced a single speech sound. I turn now to the sequential issue: Why aren't subjects able to accurately locate the cough within the sequence of phonemes? Why isn't the cough perceived in its true (isomorphic) position in the sequence? Why does the cough sound sequentially independent of the phonemes of the word, as if coexisting in a separate perceptual space (Warren & Warren, 1970)?

In the node structure theory, the cough (∗) is represented by content nodes that are unconnected to the speech perception nodes – there are no content nodes and serial-order rules for representing the vowel group, *i∗*, syllable, *gi∗*, word, *legi∗lature*, or lexical concept, *legi∗lature*. Even though speech and nonspeech noises share the same basilar membrane, perceiving nonspeech noises involves separate content and sequence nodes in an independent perceptual system, analogous in some ways to a separate sensory system. This explains why the cough (∗) is poorly localized with respect to the speech sounds, and why (in part) the cough seems to coexist in the separate perceptual space from the sentence. It also explains why speech sounds replaced by silence do not become restored. Silence is an acoustic feature of stops and becomes perceived as a speech sound in sequence with other speech sounds.

Effects of Practice

How can we perceive the serial order of sounds in words such as *sand* at rates of 20 ms per speech sound, whereas we require over 200 ms per sound for determining the order of unfamiliar sound sequences, such as *hiss-vowel-buzz-tone* (when recycled via tape loop)? This order-of-magnitude difference reflects an effect of practice under the node structure theory. Perceiving the sequence of speech sounds in *sand* depends on prior establishment of underlying nodes for the concept *sand*; the word *sand*; the stressed syllable *sand*; the initial consonant group *s*; the vowel group *and*; and the final consonant group *nd*—all of which constrain the perception of *sand* and conspire against perception of, say, *nsad* (for which there are no existing initial consonant group and syllable nodes corresponding to *ns* and *nsad*). By contrast, no underlying nodes for subsequences such as *hiss-vowel* or *buzz-tone* have been formed to constrain perception of a never previously encountered sequence, such as *hiss-vowel-buzz-tone*.

This view also captures Warren's (1974) demonstration that practice enables sequential identification of previously unfamiliar nonspeech sequences such as *hiss-vowel-buzz-tone* with durations of less than 20 ms per sound. The reason under the node structure theory is that practice enables formation of hierarchically organized higher level nodes, each representing what Warren (1974, p. 253) terms a "temporal compound," an aggregate or cluster of auditory items that is distinct from all other clusters. More specifically, nodes representing the sequence (*hiss-vowel-buzz-tone*) become connected to superordinate nodes representing, for example, a *hiss-vowel*, and a *buzz-tone*, which in turn become connected to a *hiss-vowel-buzz-tone* node. Once such mental nodes have been formed and extensively practiced, they can be activated by their corresponding sequence nodes, even with brief and recycling stimuli, because rate of priming and activation increases as a function of practice (D. G. MacKay, 1982).

Other Sequential Effects

The present chapter has only touched on some of the sequential effects that have been reported in the speech perception literature. There are others, none of which are in conflict with the node structure theory. One example is the effect of phonotactic rules or sequential constraints on phoneme identification (Massaro & Cohen, 1983b; McClelland & Elman, 1986). Another example is the fact that segment changes in "experimentally mispronounced" words are easier to detect at the beginning of a word than at the end (Marslen-Wilson & Welsh, 1978). This finding reflects the fact that prior to applying a lexical activating mechanism, priming from word-initial segments has more time to summate than priming from word-final segments. Word-initial "mispronunciations" will therefore contribute more to the total priming summated from all sources that a lexical node receives and so will play a bigger role in determining which lexical node receives most priming and gets activated when the activating mechanism is applied.

5
Temporal Organization of Perception and Action

> The production of rhythm in general is made possible by the preparation of movements and their coordination. . . . Without preparation, performance is a series of discrete reactions to external events. The development of motor skill can be traced as the progress from reactive movement to movement fluency, coupled with a flexibility in tailoring action to the details of an infinite variety of contingencies.
>
> (Shaffer, 1982, p. 110)

Theories of perception and action must account for three basic aspects of skilled behavior, and so far, the book has examined only the first two: What components underlie the organization of everyday action and perception, and what mechanisms activate these components in proper sequence?

I turn now to the third and most frequently neglected theoretical problem, that is, timing in perception and action. How do we produce behaviors with components of different durations, and how do we produce these behaviors at different rates and with different rhythms or patterns of durations? I begin with some general phenomena or constraints that any theory of timing must address, and then construct a node structure theory of timing that incorporates these general constraints and makes predictions for future test.

Requirements for a Viable Theory of Timing

The preceding chapter already has noted two important requirements for theories of timing: independence of mechanisms for timing and sequencing and a closer relationship between mechanisms for timing and sequencing than between mechanisms for timing and representing the form or content of behavior. In addition, theories of timing must address six other fundamental issues discussed in this chapter: Where in the specification of output components is rate and timing determined? What mechanisms underlie the production of rhythmic outputs? How and why is periodicity or near-miss periodicity achieved in complex skills such as typing? What is the relationship between the timing mechanisms for perception and action? What accounts for our ability to flexibly adjust the rate and timing of behavior? Why and how do actions with different timing characteristics interact with one another?

The Distributed Nature of Timing and Sequencing

The most fundamental constraint on theories of timing is that timing is an "everywhere" or distributed characteristic. Each and every component, from the lowest level nodes controlling muscle movements to the highest level nodes representing, say, sentential concepts, must be activated at some rate and for some duration. Rhythm, rate, and timing permeate the entire process of producing well-practiced behaviors, and cannot be tacked on as an independent stage at some point in the theoretical specification of the output.

To see why timing must be distributed in this sense, it is only necessary to examine existing "stage of processing" proposals (including one of my own), where rhythm and timing are treated as an "afterthought," a late stage of processing introduced just before or during the programming of muscle movements. Consider, for one example, my stage-of-processing theory (D. G. MacKay, 1971) of timing in speech production (for other examples, see Fromkin, 1973; Shaffer, 1982). Under my proposal (D. G. MacKay, 1971), the entire syntax, semantics, and phonology for producing a sentence are first constructed and then recorded within a simultaneous store or nontemporal spatial display called the phonological buffer. Only then are timing and rhythm added as part of the output specifications in the simultaneously displayed phonological string.

Stage-of-processing proposals, such as this one, face many unsolved problems. One is the complexity and reduplication of information that is required for the simultaneous store (phonological buffer). Rhythm and timing depend on information associated with units at every level (sentences, phrases, words, syllables, and segments), and the proposed spatial display must incorporate all of this information before timing specifications can be added. Because these same specifications are also needed in order to construct the sentence in the first place, adding timing after rather than during construction duplicates the specification process.

An even more serious problem for theories that represent timing as a stage of processing is speed–accuracy trade-off, one of the most pervasive phenomena in the study of skilled behavior. For all known skills, increased speed leads to increased errors in activating components, whether lower level muscle movement components, or higher level mental components. The muscle movement level has been the source of most demonstrations and theories of speed–accuracy trade-off (Meyer, Smith, & Wright, 1982), but speed–accuracy trade-off has also been demonstrated for higher level mental components. For example, consider speech errors such as the substitution of "coat-thrutting" for "throat-cutting." Here components within the phonological system have become interchanged (Dell, 1980), and in a study of experimentally induced speech errors, D. G. MacKay (1971) demonstrated that such errors increase as a function of speech rate (see also Dell, 1985b). This finding cannot be explained if rate is determined *after* specification or misspecification of phonology. For rate to influence phonological errors, phonological components and their rate of output must have received simultaneous specification. Because rate also influences errors at still higher levels, and at lower levels as well, the same argument against stage-of-processing theories applies at all levels. Instead of being specified at only one stage or level, rate and timing must permeate the entire process of producing well-practiced behavior, from the lowest level muscle movements to the highest level concepts for guiding behavior (see also Fowler's discussion of "intrinsic" timing, 1977).

By a similar argument, sequencing must also be a distributed process. The sequence of components cannot become specified at some particular stage-of-processing, because sequential errors occur at all levels of the hierarchy, from distinctive features to phrases (see Fromkin, 1973; D. G. MacKay, 1978). If the "sequencing stage" followed specification of phonemes, for example, it would be difficult to explain sequential errors that occur at lower levels, involving distinctive features, or at higher levels, involving words and phrases. Sequencing cannot be localized as a stage-of-processing at only one point in the hierarchy. Theories must represent both timing and sequencing as distributed or everywhere processes.

Monitoring, Rate, and Errors

A possible counterargument in favor of stage-of-processing theories of timing (attributable to Welford, 1968) is that errors increase with rate not because rate is a distributed characteristic but because various output-monitoring devices become suspended at faster rates, thereby allowing more errors. To be taken seriously, this explanation of speed–accuracy trade-off requires a great deal more theoretical specification and empirical support. There currently exists no empirical evidence for monitoring devices that are independent of the output mechanisms themselves (Chapter 9) and no evidence that these hypothetical output-monitoring devices are suspendable. More importantly, the concept of perceptual error monitoring as a final (but suspendable) production stage has difficulty with the time characteristics of error detection and correction. Error correction is sometimes so rapid as to precede the full-blown appearance of an error in the surface output (Levelt, 1984; and Chapter 9). Of course, monitoring and production might be considered parallel rather than strictly serial processes, but then the monitoring counterargument runs into a new difficulty. If perception proceeds in parallel with production, suspending perceptual monitors could not help production rate unless production proceeds much faster than perception. Unfortunately for the monitoring counterargument, the opposite is true; perception in fact proceeds much faster than production (Chapter 6).

The Generation of Periodicity

Although not all behavior can be described as periodic, people can and do generate perfect or nearly perfect periodicity in various ways that must be explained in theories of timing. Wing (1980) provided the most straightforward evidence for near-miss periodicity in a series of experiments using the "key-press continuation paradigm." Wing first had subjects practice tapping a key in time with the beat of a metronome and then turned off the metronome, so that they had to continue the rhythm as accurately as possible on their own.

Wing's (1980) subjects were remarkably successful in reproducing the input rhythms. Deviations from the beat were quite small and tended to alternate with one another, undershoot followed by overshoot and vice versa. Wing showed that these departures from perfectly periodic responding were attributable to two factors: imprecision in an internal clock and temporal "noise" in executing motor responses triggered by the clock. Wing also showed that both of these sources of variation appear not only in the movement itself but in the electromyographic activity that precedes onset of muscular movement, so that deviations from perfect periodicity are not attributable to physically induced lags following onset of movement.

Proficient performances of everyday skills such as Morse code, piano playing, typing, speech production, and handwriting also exhibit periodicity to varying degrees. In speech, zig–zag alternations, resembling those seen in rhythmic finger tapping, have been observed in the durations of successive phonemes and successive syllables in an utterance (Kozhevnikov & Chistovich, 1965). Periodicity and systematic (zig–zag) deviations from periodicity are also seen in typing (Shaffer, 1978) and in handwriting (Wing, 1978). The function of these periodicities in typing and handwriting is especially curious, because neither skill seems to benefit from rhythmic output. The goal of typing and handwriting is simply to produce letters as quickly as possible for a given error criterion. Apparently people not only *can*, but normally *do* generate near-miss periodicity, even when the skill requires neither rhythm nor precisely timed output.

Timing Interactions Between Perception and Production

Interactions between timing processes for perception versus action impose fundamental constraints on theories of timing, and three types of interactions have been observed: facilitative interactions, disruptive interactions, and temporal couplings between perception and action.

Facilitative Interactions Between Perception and Action

Rhythmic timing in speech production appears to facilitate speech perception. The effects on perception of sentences produced with (approximately) stress-timed rhythm provide an example. Stress-timed rhythm means that for a given rate of speech, the interval between stressed syllables in a phrase or sentence is relatively constant. For example, speakers of English produce the phrases "ran lickety split" and "the black horse" in about the same amount of time in the sentence "The black horse ran lickety split across the field," despite the two extra syllables in *lickety* (H. H. Clark & E. V. Clark, 1977; Lehiste, 1970). Considering all sentences of English, this regularity is only approximate (Fowler & Tassinary, 1981), but *producing* stress-timed rhythm could in principle facilitate *perception*, making it possible for listeners to adopt a strategy of listening most carefully and effectively at times when stressed syllables occur. And this strategy would pay off, because stressed syllables generally contain more phonetic and phonological information than unstressed syllables (Huttenlocker & Zue, 1983).

Several sources of evidence indicate that people do, in fact, expect stressed syllables at particular points in the rhythm of a sentence. Huggins (1972) is one example. By shortening one segment in a word and prolonging another, Huggins created sentences containing minor timing irregularities which subjects attempted to detect in a task requiring discrimination between these and other sentences without timing irregularities. The results showed that irregularities were more detectable when they altered the interval between stressed syllables in a sentence than when they left this interval unchanged. Changing the time between stressed syllables apparently violates the timing expectations that listeners impose on an input.

A phoneme monitoring experiment by Shields, McHugh, and Martin (1974) further illustrates the timing expectations associated with stressed syllables. Subjects pressed a key as quickly as possible after hearing the phoneme /b/ in sentences such as "You will have to curtail any morning sightseeing plans, as the plane to BENkik leaves at noon." The target sound, /b/, began a two-syllable nonsense word that received stress on either the first or the second syllable (BENkik versus benKIK) and was detected faster in the stressed syllable. However, the critical factor was the relation of the stressed syllable to the ongoing rhythm of the sentence. No difference between stressed versus unstressed syllables was found when the same words were presented in a list of nonsense syllables without sentential rhythm. People apparently listen for stress-timed rhythm only when processing real sentences.

The importance of stressed syllables in perception of English sentences appears yet again in an experiment by Allen (1972). Subjects repeatedly listened to a recording of a spoken sentence and were then asked to tap a key in time with its rhythm. That subjects were able to respond at all to this instruction is remarkable, and that they responded consistently by tapping in time with the stressed syllables is even more remarkable. It is as if stressed syllables provided the beat for an internal clock that controlled both the finger taps and perception of sentential rhythm.

Disruptive Interactions Between Perception and Action

Wing (1980) demonstrated a disruptive interaction between perceptual and motoric timing that also calls for an explanation in terms of a common timing mechanism for perception and action. Wing's subjects first learned to generate a motoric rhythm, using the key-press continuation paradigm discussed previously. In this study, however, a brief tone sounded during each press, and unknown to the subjects, the tone generator was programmed to delay or advance the tone by about 10 to 50 ms relative to the normal time when a randomly selected tap was to be made. For delayed tones, subjects inadvertently tended to delay the next one or two taps in the same direction as the tone, but not as much as would be expected if feedback determined the timing of subsequent responses (Keele, 1986), and advanced tones had no significant effect whatsoever. Auditory feedback can therefore influence the internal clock that generates finger-tapping responses but not in the manner of feedback control theory (see also Chapter 10).

The question is whether Wing's (1980) effect is related to feedback per se, or whether *any* auditory input would have a similar effect. This question was answered by Pokorny (1985), who demonstrated a similar interaction between perceptual and motoric timing by simply presenting a tone, unrelated to ongoing feedback, during the time that subjects attempted to generate a series of equal-interval finger taps. Occurrence of the tone tended to increase the intertap interval in which it occurred, even when subjects were instructed to ignore the tones. Pokorny (1985) also noted that the tone interacted with the timing mechanism, rather than with the mechanism for producing any particular response, because subsequent responses occurred at the appropriate times, rather than compensating for the response that had been mistimed. It was as if the timer were a resettable rather than persistent pacemaker. Occurrence of the tone simply reset the timing mechanism, which then continued to generate its preprogrammed equal intervals (see also Keele, 1987).

Temporal Couplings Between Perception and Action

Lashley (1951) reported an interesting interaction between input and output rhythms that likewise suggests that perception and action may share some of the same timing mechanisms. Lashley observed temporal couplings of a perceptual rhythm (listening to a band) with ongoing motoric activities such as walking, breathing, and speaking. Specifically, when someone is listening to a salient rhythm such as a marching band, the perceptual rhythm tends to cause the listener to fall in step, gesture, breathe, and even speak in time with the band. Such an interaction suggests that the same timing mechanisms may govern both the perceptual processes underlying listening to the band and the motoric processes underlying walking, breathing, and speaking (see also Keele, 1987).

Learned and Unlearned Aspects of Timing

Some aspects of timing are unlearned, and other aspects are learned, and both must be explained in theories of timing. For example, unlearned peripheral characteristics of the speech organs clearly influence timing. Because the jaw must move farther to produce an open vowel, open vowels have systematically longer durations than closed vowels (Lehiste, 1970). Differences between languages, on the other hand, illustrate a learned aspect of timing. An example is the more varied use of durational information in so-called "syllable-timed" languages such as French as compared with "stress-timed" languages such as English (Cutler, Mehler, Norris, & Segui, 1983).

Constant Relative Timing

As the overall time to produce a behavioral sequence changes due, say, to an increase in rate, the proportion of time required to produce some segment of the sequence remains constant within wide limits. This phenomenon, known as *constant relative timing*, has been observed for a large number of behaviors (e.g., walking, running, typing, handwriting, speech, lever rotation) and can be considered a general law of behavior. As a single example of this general law, consider the work of Shapiro, Zernicke, Gregor, and Diestel (1981). Subjects walked at various speeds on a treadmill, and the proportion of time required to execute the four basic phases of a step (lift, stride, heel contact, and support) was found to remain relatively invariant at the different speeds. If lift phase of a step took up 20% of step duration at a slow rate, it required about 20% at a faster rate. Timing only remained constant within limits, however. When the treadmill was accelerated beyond a certain point, for example, subjects broke into a jog, and the temporal configuration of their behavior suddenly and dramatically changed. Walking and running clearly have different temporal characteristics and are controlled by different underlying mechanisms, which nevertheless *both* conform to the law of constant relative timing. Handwriting and transcription typing also exhibit constant relative timing. Here, changes in overall output rate have been found to scale the duration of response components in almost perfect proportion, as would occur with a change in rate of a low-level internal clock (Shaffer, 1978).

Interestingly, constant relative timing has also been observed for the *involuntary* changes in the rate of behavior that occur as a result of practice. Components in a behavior sequence automatically speed up as a result of practice, and these changes in relative duration sometimes exhibit constant relative timing as well. For example, D. G. MacKay and Bowman (1969) had subjects practice producing a sentence as quickly as possible, and found that they produced the sentence faster after 12 trials of practice than after only one. More importantly, different components of the sentence speeded up proportionally; the relative durations of words and of syllables remained constant at the fastest speed. If a word took up 10% of total sentence duration at the slower, less practiced rate, then it took up about 10% of sentence duration at the faster, more practiced rate (see also D. G. MacKay, 1974). The constant relative timing that occurs when behavior speeds up, either voluntarily or involuntarily (as a result of practice), places fundamental constraints on theories of timing.

Theories postulating an on-line process of calculation for altering durations of behavioral components produced at different rates (e.g., Shapiro et al., 1981) have difficulty explaining constant relative timing, because the phenomenon appears in the behavior of insects and crustaceans, where such calculations are unlikely. Constant relative timing also appears immediately after subjects change their rate, without the lag times that seem necessary for temporal calculations. Rate-dependent changes in timing call for a more automatic mechanism that does not recompute movement time on the basis of rate.

Deviations From Constant Relative Timing

Constant relative timing cannot be expected for all response components. In particular, not all changes in speech rate can be expected to scale proportionally over the durations of vowels versus consonants. With voluntary changes in speaking rate, vowels exhibit much more "elasticity" than do consonants. Vowels can be prolonged almost indefinitely to slow down the rate of speech, but a greatly prolonged stop consonant no longer resembles speech.

Such observations suggest that two different timing mechanisms may control production of speech sounds, one for consonants, the other for vowels. Consistent with this hypothesis, Tuller, Kelso, and Harris (1982) found that the lag with which a consonant is initiated following a vowel remained constant at different rates of speech, but only when compared to the interval between vowels. Neither consonant nor vowel duration per se remained constant relative to overall utterance duration.

Temporal Interactions

Theories of timing must explain why concurrent actions with different timing characteristics tend to interfere with one another. When unpracticed subjects attempt to produce several actions at once, they experience considerable difficulty when concurrent movements conflict in timing, but little difficulty when concurrent movements are temporally compatible or occur at harmonically related times. A study by Klapp (1979) provides a clear demonstration of this effect of temporal compatibility on the ability to time concurrent activities. Subjects pressed telegraph keys, one for each hand, in time with tones presented periodically to the two ears via headphones. The goal was to maximize temporal overlap of the key press and the tone to the corresponding ear.

Tones arriving at the two ears were either temporally compatible or incompatible. In the temporal compatibility condition, rhythms to the two ears were synchronous and harmonically related. One series proceeded at twice the rate of the other. In the temporally incompatible condition, rhythms to the two ears were equally fast on the average but were desynchronized, that is, they occurred at harmonically unrelated times. The results were straightforward; average temporal overlap of tone and key press was greater when the two rhythms were temporally compatible than when they were temporally incompatible (Klapp, 1979).

Klapp (1981) extended this finding by showing that temporal incompatibility disrupts concurrent actions with two quite different motor systems: speech and finger movement. The subjects simultaneously tapped telegraph keys and prouced syllables in time with corresponding perceptual rhythms that were either temporally compatible or temporally incompatible with one another. As before, the greatest disruption occurred with temporally incompatible inputs, as if the same internal clock were being used for timing both speech and hand movements.

Effects of Practice

Effects of practice are everywhere apparent in the timing literature and must be explained in theories of timing. For example, in skills that do not require rhythmic timing, such as typing, near-miss periodicity only becomes apparent when the skill has been highly practiced. Genest (1956) found that intervals between typestrokes came closer and closer to perfect periodicity as typists became progressively more proficient, but observed no periodicity whatsoever during early stages of learing to type. Periodicity apparently emerges as a function of practice.

The periodic timing of keystrokes becomes especially obvious when highly skilled typists transcribe experimentally constructed materials known as "alternation passages." These passages contain phrases, such as "authentic divisors," where normal typing conventions require alternate hands for each stroke, thereby minimizing interactions between keystrokes with fingers of the same hand. In typing these passages, the interkey intervals of expert typists become nearly equal, and subsequent strokes tend to compensate for deviations from perfect periodicity. That is, an especially fast stroke tends to follow, and make up for, an especially slow one and vice versa. As Shaffer (1980, p. 116) points out, this zig–zag, or negative serial covariance, sometimes approaches the theoretical limit that could be expected for a perfectly periodic internal clock.

Practice also plays a role in effects of temporal incompatibility (discussed previously). The original effects of temporal incompatibility (Kelso, Southard, & Goodman, 1979; Klapp, 1979; 1981) were obtained with unpracticed subjects, and it is clear that skilled performers can learn to produce temporally incompatible activities. For example, Lashley (1951) observed that expert pianists can readily produce temporally incompatible rhythms, such as a 3/4 rhythm with one hand (3 beats per measure) and a 4/4 rhythm with the other. Shaffer (1980) likewise observed that concert pianists could produce temporally incompatible finger movements by shifting one hand off the beat maintained by the other, for example. It is as if practice enables each hand to become controlled by its own internal clock, or by independent pulses from a single internal clock with a very rapid pulse rate, so that every Nth pulse triggers the program for one hand, and every Mth pulse triggers the program for the other.

The speedups in behavior that themselves result from practice also exhibit practice effects (D. G. MacKay & Bowman, 1969). If components of a behavior sequence have received relatively equal or asymptotic degrees of prior practice, as was the case for the phonemes that subjects in D. G. MacKay and Bowman (1969) produced, then proportional or constant relative speedups in these components can be observed as a result of practice. However, when different components of a behavior sequence have had unequal and nonasymptotic levels of prior practice, unpracticed components speed up faster than practiced components, so that relative timing changes. Consider for example the findings of Seymour (1959), von Treba and Smith (1952), and Wehrkamp and Smith (1952). Subjects practiced making a three-component response in the following sequence: (1) grasp a small object, (2) move it over to a box, and (3) drop it into the box as quickly as possible. With errors held constant, the time required to move the object improved faster with practice than did the time required for grasping it and dropping it into the box. The reason is related to prior practice. The actions of grasping and releasing small objects have had extensive prior practice, and so benefit little from further practice. In contrast, the action of moving the arm with a particular load, in a particular direction, and over a particular distance has received relatively little practice, and so improves quickly with additional practice.

The Node Structure Theory of Timing in Perception and Action

As developed so far, the node structure theory postulates a hierarchy of content nodes for representing the form of perception and action, while an independently stored set of (sequence) nodes codes serial-order rules and determines the sequence in which content nodes become activated. I turn now to timing nodes: their structural organization, rationale, and functioning; how they interact with one another; when their connections become established; and how they generate near-miss periodicity. I then focus on applications of the theory to phenomena such as constant relative timing. I conclude with evidence for a central thesis of the node structure theory, that one and the same timing node can play a role in both perception and action.

The node structure theory of timing was explicitly intended to meet the constraints on theories of timing discussed previously. Timing in the node structure theory is an inherent part of all output specifications, from the lowest level nodes controlling muscle movements to the highest level nodes representing sentential concepts, and is therefore a distributed or everywhere characteristic. Rhythm, rate, and time permeate the entire process of producing well-practiced behaviors and are not tacked on as an independent stage at some point in the specification of output components. Rather, each and every content node in an output hierarchy is activated at some rate, with some periodicity, and with some duration or is not activated at all. Finally, mechanisms for the timing and sequencing of behavior in the node structure theory are independent but closely related. In other words, separate node systems in the theory represent the form, sequence, and timing of behavior, but timing nodes are more closely related to the sequence nodes than they are to content nodes for representing the form of behavior. This relationship occurs because timing nodes have direct or first-order connections with sequence nodes but not with content nodes.

Structural Characteristics of Timing Nodes

A timing node is connected to and activates a set of sequence nodes that can be seen to fit the definition of a domain. For example, the sentential timing node is connected to the domain of sentential sequences nodes, and the phonological timing node is connected to the domain of phonological sequence nodes. Because all nodes are part of some domain, domains can therefore be considered a distributed rather than a local characteristic of node structures.

Timing nodes are connected with and activate sequence nodes in the same way that sequence nodes are connected with and activate content nodes, that is, via the most-primed-wins principle. When a timing node becomes activated, it multiplies the priming of the domain of sequence nodes connected to it until the most primed sequence node reaches threshold and becomes activated. By determining when the sequence nodes become activated, timing nodes therefore determine the temporal organization of the output.

Timing nodes play an essential role in the organization of content and sequence nodes into systems, and indeed, can be considered a defining characteristic of automatized systems. An automatized system is a set of nodes that becomes activated by means of a unique set of timing nodes. Timing, sequence, and content nodes bear a hierarchic relationship to one another in the activation of skilled behavior. Each timing node is connected with a domain of sequence nodes, and each sequence node is connected with a domain of content nodes. A single timing node can therefore be said to activate an entire system of sequence and content nodes at some particular rate. As already noted, for example, at least three timing nodes are required for producing speech at any given rate, the *sentence time* node, the *phonological time* node, and the *muscle time* node. The sentence time node is connected to the dozens of sequence nodes representing sequential rules for English sentences. The phonological time node is connected to the dozens of sequence nodes representing sequential rules for English phonology. And the muscle time node is connected to the dozens of sequence nodes for sequencing muscle movements in producing English speech sounds.

The relation between timing nodes and content nodes is always indirect, always one-to-many, and sometimes also many-to-one. Timing nodes always have a one-to-many relationship to content nodes, because each timing node is connected to many sequence nodes, each of which serves (indirectly) to activate many content nodes. However, some timing nodes also have a one-to-many relation to content nodes, because several timing nodes can serve (indirectly) to activate the same content node. Consider the muscle movement nodes that move the arm, for example. One set of timing nodes controls the arm during reaching, whereas a different set of timing nodes controls the coordination of the arms (and legs) during walking.

The Rate of Perception and Action

Timing nodes determine the rate at which perception and action can occur. Different timing nodes generate different periodicities or average rates of activation. For example, the sentential timing node exhibits a slower periodicity than the timing node for the speech muscles, because muscle flexions and extensions are produced orders of magnitude faster than words and other sentential components. The periodicity of a timing node determines the overall rate of perception or action. Consider speech rate, for example. To determine a desired rate of speech, speakers need only adjust (voluntarily) the overall periodicity or pulse rate of the relevant timing nodes (e.g., fast, normal, or slow).

As in production, the actual rate setting of the timing nodes in perception is partly individual specific, and partly situation specific, determined by the perceived rate of input, for example. This enables the input, and perception of the input, to take place at different rates within wide limits (see also Chapter 4). For example, rate settings for the timing nodes of speaker and listener need not match for speech perception to occur in the node structure theory. The only constraint is that the perceiver's rate setting not be so slow that priming has decayed to below commitment threshold by the time that the next pulse from the timing node arrives, and not so fast that so little priming has built up that the probability of activating the wrong node exceeds the error criterion (D. G. MacKay, 1982).

The Engagement and Disengagement of Timing Nodes

The way that timing nodes are connected to sequence nodes enables simple higher level decisions regarding timing nodes to selectively engage or disengage whole systems of content nodes. Internal rather than overt speech is produced, for example, by engaging the timing nodes for the sentential and phonological systems and disengaging those for the muscle movement system. As a consequence, phonological components for a sentence become activated in proper serial order and prime their corresponding muscle movement nodes (D. G. MacKay, 1981), but full-fledged movement of the speech musculature does not ensue. No content nodes for muscle movement can become activated, because no timing and sequence nodes within the muscle movement system have been activated.

A high-level decision is required to determine which timing nodes are to become engaged and when. How fast an action unfolds, for example, is determined by altering the periodicity of a timing node or by selectively engaging timing nodes with different periodicities. For example, a given system may have three timing nodes, one each for fast, medium, and slow rates. In the case of hierarchically organized systems—say, the sentential, phonological, and muscle movement systems for speech—the timing nodes corresponding to a given rate are coupled, fast with fast, medium with medium, and slow with slow. As a result, only a simple decision (fast, medium, or slow) must be transmitted to one of the systems, in order to set the rates for the remaining systems. Beyond being a single and simple decision, however, little is known about how such decisions are represented and how they are executed. We do know that some perceptual events such as the unexpected occurrence of an auditory tone during an equal-interval finger-tapping task can cause resetting or reengagement of a timing node, so that the intervals generated begin again from the tone (Pokorny, 1985).

In this sense, timing nodes are flexible or resettable pacemakers rather than persistent pacemakers (Keele, 1986). Once activated, timing nodes don't simply continue activating sequence nodes indefinitely into the future, uninfluenced by external events.

Selective engagement of different sets of timing nodes can also determine the mode or modality of output. Consider, for example, two common modes of speech production: overt versus whispered speech. Overt speech engages all of the timing nodes for the sentential, phonological, and muscle movement systems. Whispering engages exactly these same timing nodes, except for the muscle movement timing nodes that control voicing within the larnyx. Selectively neglecting to activate this one timing node devoices the output, as during whispering (for further details, see D. G. MacKay & MacDonald, 1984).

The ability of bilinguals to rapidly alternate between languages illustrates another switch in output modality that could be achieved by selectively engaging timing nodes. Moderately proficient bilinguals require only about 0.2 s to switch between languages (Kolers, 1968), and fluent bilinguals, for example, French Canadian broadcasters, can switch languages with virtually no lags whatsoever. They produce the sentence "In one corner of the room stood three young men" about as fast as the mixed, French–English sentence "Dans une corner of the salle stood trois jeunes hommes."

How can the phonological systems for different languages become engaged so quickly? Content nodes representing such a sentence within the sentential system are identical for both languages (D. G. MacKay & Bowman, 1969), but phonological nodes for the two languages differ. This means that proficient bilinguals can automatically produce words in first one language and then the other by engaging the phonological timing nodes for the two languages in alternation. Thus, activating a lexical content node primes phonological nodes in both language systems, and alternate engagement of the phonological timing node for English and then French in rapid succession automatically causes production of words in first one language and then the other.

Of course, switching between different output modes only becomes necessary when both modes require simultaneous and incompatible use of identical muscles. The different muscles required for compatible output modes *can* be activated simultaneously. For example, English and American sign language (Ameslan) are compatible output modes for many sentences, involving identical sentential nodes but different muscle movement nodes. Fluent English–Ameslan bilinguals can simultaneously produce many sentences in both languages without interference—in theory, by simultaneously engaging the timing nodes for both output modes.

Practice, Automaticity, and Periodicity

Timing nodes represent the last stage in the establishment of automaticity in the node structure theory (D. G. MacKay, 1987). Timing nodes cannot be used to control behavior during early stages of skill acquisition when new connections between content and sequence nodes are being formed. The theory therefore predicts that timing periodicities will only become apparent when a skill has been highly practiced and is automatic. Genest's (1956) observations on periodicity in typing are a case in point. Periodicity was observed during the later stages of learning to type but not during the earlier stages.

Other Effects of Practice

The original temporal compatibility effects of Kelso et al. (1979) and Klapp (1979; 1981) were obtained with unpracticed subjects, and the node structure theory predicts that skilled performers can learn to produce temporally incompatible activities with prolonged practice. Practice enables the expert to develop independent timing nodes for controlling temporally incompatible activities. As a result, concurrent activities are no longer constrained to begin and end at the same time, as would be the case if the activities shared a single timing node. This explains Lashley's (1951) observation that expert pianists can produce without interference a 3/4 rhythm with one hand (3 beats per measure), and a 4/4 rhythm with the other. It also explains Shaffer's (1980) observation that concert pianists can shift one hand off the beat maintained by the other. It is as if independent internal clocks have been developed to control each hand. Time-sharing is another possibility. For example, a single internal clock may be emitting a very rapid pulse rate, so that, with the help of a counter, every Nth pulse executes the program (i.e., activates the most primed nodes) for one hand, and every Mth pulse executes the program for the other. As D. G. MacKay (1985) points out, counters are theoretically necessary for explaining other aspects of timing behavior, and counters enable a pacemaker timing node to meter out intervals of (almost) any length, thereby acting as an interval timer in the sense of Keele (1986; 1987).

The Coupling of Timing Nodes

Unlike sequence nodes, different timing nodes can be coupled so as to emit pulses in unison. These couplings enable simultaneous control and synchronization of a large number of systems for perception and action. In this way, a single timing node, hierarchically coupled to several subordinate timing nodes, can indirectly activate nodes at regular time intervals, within either a perceptual system, an action system, or several perception–action systems that are simultaneously generating coordinated movements. As a result, evidence for periodicity in a perceptual system, an action system, or several simultaneously coordinated perception–action systems can be construed as evidence for timing nodes.

Coupled timing nodes provide a mechanism whereby the activity of different output systems can become coordinated. The precise and naturally occurring synchronizations of speech and gesture that occur when people simultaneously refer to and point toward an object (Levelt, Richardson, & Heij, 1985) illustrate how coupled timing nodes can coordinate different output modalities. The coordinated activity of the three systems for producing overt speech illustrate the hierarchic coupling of timing nodes within the same modality. The sentence, phonological, and muscle movement timing nodes have different average pulse rates. The *phonological time* node generates more pulses per second than the *sentence time* node, because phonemes are produced faster than words (by a factor of about 5 on the average). The muscle time node generates even more pulses per second, because muscle movements are produced faster than phonemes and words. However, the three timing nodes must operate in conjunction. If the *sentence time* node is speeded up, the *phonological time* and *muscle time* nodes must be speeded up proportionally. If a higher level timing node becomes decoupled and generates pulses too quickly, relative to timing nodes at lower levels, behavior will break down, and large numbers of errors will occur. With the possible exception of "cluttering," a close relative of pathological stuttering, catastrophic malfunctions of this sort seem to be rare, and one wonders what mechanisms have evolved to prevent them. Perhaps a built-in coupling connection prevents the timing nodes for different naturally coordinated systems from becoming desynchronized in this way. However, not all couplings between timing nodes are built-in. To illustrate an acquired (and unfortunately rather complex) coupling between timing nodes, I examine the phenomenon of (approximately) stress-timed rhythm in greater detail in the following section.

Stress-Timed Rhythm and Coupled Timing Nodes

Stress-timed rhythm means that for a given rate of speech, the interval between stressed syllables tends to remain relatively constant. We require about the same amount of time to produce the words *summertime* and *mealtime* in the sentence "Mealtime differs in the summertime," despite the extra syllables and segments in *summertime*. The node structure theory tolerates two different accounts of the production of stress-timed rhythm. Both accounts require coupled timing nodes for activating stressed and unstressed syllables, and a counter, which is a mechanism that can both count the number of pulses that a timing node generates and engage or disengage the timing node. (For empirical and theoretical arguments in support of counters, see D. G. MacKay, 1985.)

The more complex account of stress-timed rhythm is this. The sequence nodes for the domain (stressed syllable) are connected to a stressed-syllable timing node, which causes stressed syllables to become activated at regular intervals. The timing of unstressed syllables introduces the complexity. Sequence nodes for the domain of unstressed syllable nodes are connected with the counter, plus a set of at least four timing nodes. These four unstressed-syllable timing nodes are coupled with the stressed-syllable timing node, but have a periodicity that is 2, 3, 4, or 5 times as fast. Once a stressed syllable becomes activated, the counter registers how many unstressed syllable nodes have been primed or readied for activation prior to the next stressed syllable. The counter then engages the appropriate unstressed-syllable timing node for generating either 1, 2, 3, or 4 pulses, so that the interval between stressed syllables remains equal.

The simpler explanation of stressed-timed rhythm is that a single clock and counter controls both stressed and unstressed syllables. The counter counts the pulses and causes activation of a stressed syllable on every sixth beat, say, while causing activation of unstressed syllables on the five intervening beats. However, the stressed syllable becomes lengthened in the surface output, depending on the number of unstressed syllables primed to be activated during the intervening interval.

Deviations From Periodicity

Because timing nodes for skilled behavior have their own endogenous rhythm and become active at regular intervals, the generation of periodicity is a general characteristic of the node structure theory, and the occurrence of perfect, or nearly perfect, periodicity in behavior and in electromyographic activity that precedes behavior (reviewed previously) is therefore to be expected. What require special attention under the theory are *deviations* from periodicity in skilled behavior. As discussed below, the node structure theory predicts deviations from periodicity under three conditions: when the skill involves multiple levels of timing control; when low-level timing factors within, say, the muscle movement system mask the periodicities of a higher level timing node; and when the skill depends on the processing of external feedback that is unpredictable in time.

Multiple Levels of Timing Control

Deviations from stress-timed rhythm in the production of English sentences illustrate how aperiodic events at one level can mask the periodic events at higher or lower levels. Measurements by Lehiste (1977), for example, show that the interval between stressed syllables may deviate by a factor of five for different sentences of English. Several factors contribute to these deviations from stress-timed rhythm under the theory. One is the time required for activating phonological and muscle movement nodes below the syllable level. For example, with lexical stress equated, more time is required to produce a seven-segment syllable, such as *splints*, than a two-segment one, such as *in*. Adding consonants to a syllable compresses the duration of each segment to some extent, but not so much as to cause invariance. The context in which a surface segment occurs can also influence its duration, because vowels become greatly lengthened when the following consonant is voiced rather than unvoiced (e.g., Lehiste, 1970). Finally, as discussed earlier in the chapter, some segments are more "compressable" than others. For example, vowel duration can change more as a function of lexical stress and overall speaking rate than can consonant duration.

How are these deviations from stress-timed rhythm to be accounted for in theories of English sentence production? Under the node structure theory, a single timing node activates nodes for stressed syllables at regular intervals, but low-level factors, such as number of segments per syllable, sequential interactions between the segments, and the temporal inflexibility of consonants, conspire to alter the periodicity of stressed syllables in the surface output. Such factors can distort a higher level (or underlying) periodicity, because an activated content node only primes its subordinate content nodes and cannot completely determine when they become activated.

Factors above the level of the syllable can also influence timing in the surface output and can distort a periodicity occurring among syllable nodes. An example is the constituent-final pause, which occurs in most, if not all languages. Speakers almost invariably insert pauses after major syntactic constituents, as in "When Mary leaves, [pause] Sam will be upset." Without this pause, listeners could readily mishear and misunderstand this sentence as, "When Mary leaves Sam, [pause] we'll be upset" (W. E. Cooper & Paccia-Cooper, 1980). Constituent-final pauses are necessary to prevent such misunderstandings and automatically override the timing periodicities of lower level units, such as syllables. As a result, demonstrating these lower level periodicities at the surface level will be difficult.

Mechanisms underlying production of constituent-final pauses raise interesting subsidiary questions. W. E. Cooper and Paccia-Cooper (1980, see also Gee & Grossjean, 1983) argue that constituent-final pauses arise from syntactic or sequential rather than semantic mechanisms, and the syntactic nature of this mechanism is readily explained under the node structure theory. Constituent-final pauses reflect activation of the sequence node SENTENTIAL PAUSE, which has effects resembling those of the sequence node PAUSE for spacing Morse code letters (discussed in D. G. MacKay, 1985). SENTENTIAL PAUSE participates in the sequential rules for sequencing phrases and sentences, and at a much higher level than the Morse code PAUSE. As would be expected under the node structure theory, SENTENTIAL PAUSE plays as much of a role in perception as in production. For example, people hear pauses at the major constituent boundaries of experimentally constructed sentences, even when no pause is actually present (Lieberman, 1963).

Low-Level Masking of Periodicity

Typing illustrates how low-level factors can mask a higher level periodicity. Indeed, low-level masking of periodicity is so common in typing that deviations from periodicity have been less of a research issue than the existence of periodicity itself (Gentner, 1983; Norman & Rumelhart, 1983). Recall that interkey intervals tend to approach equality only when highly skilled typists transcribe specially constructed materials, "alternation passages," where the hands alternate for each stroke. What is the basis for this periodicity, and why does it only become manifest with alternating keystrokes and after extended practice? The main determinant under the theory is a timing node controlling the domain of sequence nodes for activating keystrokes. The periodicity of the keystrokes reflects the periodicity of the timing node. However, this periodicity only becomes evident in proficient typists, because timing nodes only become connected during the final stages of automatization when, for example, visual input is no longer needed to control the keystrokes. Alternation passages also help to reveal the periodicity because between-hand strokes eliminate two major muscle movement factors that normally mask higher level periodicities. One factor is preparation time. When fingers of the same hand must strike two keys in succession, positioning the hand and relevant fingers requires a preparation time that varies with location of the immediately prior key. Interactions between fingers of the same hand moving toward about-to-be-typed keys are the other factor. Whenever successive keys require same-hand fingers to move in opposite directions, interstroke intervals lengthen.

Similar sources of "low-level noise" suggest an explanation for the deviations from periodicity observed by Wing (1980) in finger-tapping movements and in electromyographic activity that precedes onset of these tapping movements. Timing in typing, speech, and finger tapping are therefore similar. In all three behaviors, lower level factors introduce deviations from periodicity imposed by higher level timing nodes.

Higher level factors also play a role in masking lower level periodicities in these and other skills, and it is possible to manipulate these masking effects. In typing, for example, one can expect keystroke intervals to become aperiodic when skilled typists type freestyle, or from dictation, rather than transcribing from printed texts, as in most studies to date. Dictation and freestyle typing introduce higher level timing factors, which mask the periodicity of the timing node for activating keystrokes in a manner resembling constituent-final pauses in speech.

Applications of the Theory

To illustrate how the theory works in further detail, I now reexamine phenomena such as constant relative timing and the effects of temporal compatibility in perception and action.

Constant Relative Timing

As already noted, the phenomenon of constant relative timing calls for an automatic timing mechanism (without internal calculations for determining the temporal durations of output components produced at different rates), and the node structure theory provides such a mechanism. Constant relative timing is an automatic and emergent property in the node structure theory (D. G. MacKay, 1983). By way of simple hypothetical example, compare a behavior being produced at normal rate, with the same behavior being produced at about half that rate. To generate output at the normal rate, the timing node controlling the lowest level components of the behavior activates a sequence node and its most primed content node at regular intervals, say every n ms. Following a higher level decision to cut the output rate in half, a timing node with half the number of cycles per second becomes engaged. As a result, the same content nodes will become activated every 2n ms, and duration of each component will double. In the absence of factors such as those discussed previously that mask or distort this periodicity, the system will automatically exhibit perfect constancy, because changes in overall rate bring corresponding and proportional changes in the onset time for each component.

Data from handwriting and from transcription typing fit this surprisingly simple account remarkably well. In both skills, changes in overall rate of output have been found to scale the duration of response components in almost perfect proportion, as would occur with a change in rate of a low-level timing node. As already noted, however, changes in overall rate of speech are not scaled proportionally over the durations of all segments. In particular, Tuller et al. (1982) found that consonant durations remained constant relative to vowel durations over substantial changes in speech rate, but only when compared to the interval between vowel onsets. Neither consonant duration nor vowel duration per se remained constant relative to overall utterance duration. The reason is that vowels are much more "elastic" than consonants. Vowels can be prolonged until breath runs out, whereas stop consonants are difficult to prolong, and in any case, they no longer sound like speech when greatly prolonged. Vowels and consonants apparently engage two separate but coupled timing nodes, and although both are muscle movement timing nodes, the one for consonants has a narrower range of rate settings than the one for vowels to which it is coupled.

Consider now the way that components speed up as a result of practice. The theory predicts that behaviors will exhibit constant relative timing in the speed-ups that result from practice, but only if all segments of the behavior sequence have had roughly equal prior practice, as would be the case for the words and syllables of the sentences practiced in D. G. MacKay and Bowman (1969) (see also D. G. MacKay, 1982). However, not all behaviors will exhibit constant relative timing in the speed-ups that result from practice, because less practiced nodes benefit more from further practice than do more practiced nodes (D. G. MacKay, 1982). As a consequence, when segments of a behavior sequence have had unequal prior practice, unpracticed segments will speed up faster than practiced segments, so that relative timing will change. This explains Seymour (1959), von Treba and Smith (1952), and Wehrkamp and Smith's (1952) observations that the time required to move a small object improves faster with practice than does the time required for grasping it and dropping it into the box. Nodes for grasping and releasing small objects have had extensive prior practice and so benefit little from further practice, whereas nodes for moving the arm when carrying a particular load, in a particular direction, and over a particular distance have received relatively little practice, and so speed up a great deal with additional practice.

Temporal Compatibility in Perception and Action

To further illustrate the difficulty of generating temporally incompatible responses, consider the findings of Kelso et al. (1979). Subjects placed their index fingers on starting keys directly in front of them, and following a "go" signal, simultaneously moved both hands as quickly as possible to hit targets to the left and right of starting keys. The target for one hand was much larger than the target for the other, and according to Fitts' law (Keele, 1981), the hand for the larger target should begin to move sooner and reach its destination faster than the hand for the smaller target. However, detailed analysis of the movements showed that both hands began to move at the same time following the go signal, proceeded at the same speed, and contacted their targets simultaneously. A single timing node was apparently coordinating the simultaneous motion of the hands, so that the phases of both movements assumed the same temporal pattern as the more difficult of the two actions.

Kelso (in a related experiment cited in Keele, 1981) made another important discovery that bears on the relationship between sequencing and timing, and should apply to both skilled and unskilled performers under the node structure theory. Using the same apparatus and procedures discussed above, Kelso instructed subjects to terminate one movement before the other, with as small a difference in termination time as they could generate. These instructions had two unexpected effects. First, subjects greatly increased the time to begin movement, far beyond the generated difference in termination time. Second, contrary to instructions, subjects often desynchronized the *initiation*, as well as the termination, of their movements, starting *and* stopping one movement before the other.

Both findings are readily explained under the node structure theory. Instructions to desynchronize simultaneous movements require the formation of a serial-order rule, instantiated via inhibitory connections between sequence nodes, so as to terminate movement of, say, the right hand first, then the left. These sequence nodes must become connected to the proper content nodes; if connected in error to the content nodes controlling the entire movement, rather than just those controlling termination, then movement initiation will also become desynchronized, as was found. Moreover, activating these sequence nodes takes time, and must precede activation of the content nodes, which explains why more time was required to *begin* movement with either hand in this task.

Cognitive Clocks and Temporal Compatibility

A central thesis of the node structure theory is that higher level timing nodes for perception and action are shared. Timing nodes controlling systems of mental nodes are both perceptual and motoric. Timing nodes represent cognitive clocks, which play a role in both perception and action. As a result, the node structure theory predicts effects of temporal compatibility in perceiving input strings and, more importantly, interference between simultaneous perception and production of such strings. Some of the evidence supporting these predictions is discussed in the following section (see also Keele, 1987).

Temporal Compatibility and the Perception of Rhythm

Klapp, Hill, and Tyler (1983) were the first to demonstrate effects of temporal compatibility in perception. Subjects listened to a series of tones while watching a series of flashes arriving at times that were either temporally compatible or temporally incompatible with the tones. The task was to press a single switch as quickly as possible to indicate that either the tones or the flashes had stopped. Reaction time was faster when the rhythms were temporally compatible (harmonically related) than temporally incompatible. Because the task was to monitor rather than to produce the rhythms, this effect of temporal compatibility cannot be attributed to the production of a motor response. Rather, rhythmic compatibility must play a role in perception as well as production of rhythms. For unpracticed subjects, a single timing node apparently governs the timing expectations for both visual and auditory perception (see also Keele, 1987).

6
Asymmetries Between Perception and Action

> Many psychologists, including the present authors, have been disturbed by a theoretical vacuum between perception and action. The present volume is largely the record of prolonged—and frequently violent—conversations about how that vacuum might be filled.
>
> (Miller et al., 1960, p. 11)

> Roughly speaking, the listener has to reverse the process of speaking.
>
> (Bierwisch, 1966/1985, p. 123)

The present book has so far concentrated mainly on symmetries or similarities in the structures and processes for perception and production. Mental nodes represent the main structural symmetry, and these shared theoretical components for perception and action contribute to empirical symmetries such as the Freudian error symmetry (see also Chapter 2). The Freudian error symmetry refers to the virtual indistinguishability of Freudian misperceptions versus misproductions. Compare these examples: *carcinoma* substituted for *Barcelona* by someone preoccupied with the disease, and *battle scared general* substituted for *battle scarred general* by a speaker believing the general to be scared of battle (see Chapter 4). One error is a misperception, and the other is a misproduction, but it would be difficult for an otherwise uninformed observer to tell which was which.

The fact that perception and production share identical microprocesses (e.g., priming and activation; Chapter 1) represents the main processing symmetry in the node structure theory. This **microprocessing symmetry** contributes to empirical symmetries such as the fact that speed or rate of processing trades off with errors in the same way for both perception and production. Misperceptions and misproductions increase with rate of processing, because priming must summate over time before an appropriate node can be activated. In comprehending the word *Barcelona*, for example, its lexical content node, *Barcelona*(noun), must acquire greater priming than any other (extraneous) node in its domain when its activating mechanism is applied. Because extraneous nodes receive priming which approximates a *random* distribution at any point in time, whereas the appropriate or summating-from-below node receives priming which increases systematically over time, the appropriate node eventually acquires more priming than these extraneous nodes if priming is allowed to accumulate for long enough. Thus, reducing the time available for summation increases the likelihood of error, because errors occur when the appropriate node lacks sufficient time for its summated priming to exceed that of all other nodes in its domain when the activating mechanism is applied. This same summation process also explains the greater frequency of misperceptions on word-middle than word-initial segments (Chapter 4) as well as speed–accuracy trade-off in production (Chapter 3).

However, many phenomena are asymmetric between perception versus production, and theories must explain asymmetries as well as symmetries. The present chapter concentrates on asymmetries and shows how they arise from structural and processing asymmetries in the node structure theory. Table 6.1 provides an overview of the theoretical symmetries and asymmetries together with the empirical symmetries and asymmetries to which they contribute. To

TABLE 6.1. A summary of symmetries and asymmetries in the node structure theory and the empirical symmetries and asymmetries to which they contribute.*

Symmetries in the Node Structure Theory

I. **Mental Node Symmetry**
Freudian Symmetry
II. **Shared Microprocess Symmetry**
Speed–Accuracy Symmetry

Structural Asymmetries in the Node Structure Theory

I. **Muscle Movement Asymmetry**
Stuttering Asymmetry
II. **Sensory Analysis Asymmetry**
Masking Asymmetry
III. **The Uniqueness Asymmetry**
Sequential Domain Asymmetry
Garden Path Asymmetry
IV. **The Linkage Strength Asymmetry**

Processing Asymmetries in the Node Structure Theory

I. **The Priming Summation Asymmetry**
The Maximal Rate Asymmetry
Phonological Similarity Asymmetry
Sequential Error Asymmetry
Position-Within-a-Word Asymmetry
Word Boundary Asymmetry
The Synonymic Error Asymmetry
II. **The Connection Formation Asymmetry**
The Word Production Asymmetry
III. **The Level of Activation Asymmetry**
The Maximal Rate Asymmetry
The Lexical Error Asymmetry
IV. **The Sequential Activation Asymmetry**

*Theoretical symmetries and asymmetries are boldfaced; empirical symmetries appear in italic.

distinguish the empirical from the theoretical, I have used bold type to identify *theoretical* symmetries and asymmetries in Table 6.1 and throughout the chapter. Although some of these theoretical asymmetries have already been mentioned earlier in the book, bringing them together here not only helps to explain the empirical asymmetries, but provides a useful contrast with "strictly symmetrical" theories of relations between production and perception, discussed in the following section.

The Importance of Asymmetries Between Perception and Action

Theories incorporating identical perception–production components often assume symmetrical processes for perception and production (Bierwisch, 1966/1985). Perceptual processes in these theories are simply the reverse of corresponding production processes, like the bidirectional reactions in chemical formulas. Gordon and Meyer (1984, p. 171) use a flow chart to summarize current theories incorporating this "symmetry assumption." In the chart, arrows in one direction represent perceptual processes, while arrows in the opposite direction represent production processes.

Symmetry between perception and production has been a popular assumption (Fodor et al., 1974), which in principle enables researchers to devote all of their efforts to studying perception. If perception and production engage symmetric processes, studies of production are redundant and unnecessary. Solving the problem of perception also solves the problem of production. Like the philosophical and theoretical traditions discussed in Chapter 1, the symmetry assumption subordinates action and encourages researchers to adopt a perception-without-action approach.

The main point of the present chapter is that the symmetry assumption is incorrect. Perception and production engage identical microprocesses but in asymmetric ways. These asymmetries indicate that studies of production and comparisons between perception and production are both necessary and theoretically important. The asymmetries also call into question the philosophical traditions that consider it necessary and sufficient to study perception-without-action.

Empirical Asymmetries Between Perception Versus Production

A review of the literature reveals four general classes of empirical asymmetries between perception versus production: the word production asymmetry, the maximal rate asymmetry, the listening practice asymmetry, and asymmetries between slips of the tongue versus slips of the ear. Not all of these asymmetries have been firmly established. Some have yet to undergo statistical comparison (for methodological reasons discussed later in the chapter), but all are large in magnitude and enjoy theoretical support. As will be shown, the node structure theory predicts them all.

The Word Production Asymmetry

The word production asymmetry is the oldest and most famous of the empirical asymmetries and refers to the fact that production vocabularies tend to be much smaller than recognition vocabularies. In general, children can recognize and understand a word long before they can use it in speech production (E. Clark & Hecht, 1983).

The Maximal Rate Asymmetry

The maximal rate asymmetry is one of the most striking differences between speech perception and speech production, and refers to the fact that we can perceive speech at a faster rate than we can produce it. Foulke and Sticht (1969), Duker (1974), and Seo (1974) have summarized the evidence on the perception of speeded or time-compressed speech.

Electromechanical devices that systematically sample and compress acoustic signals provide a wide range of acceleration without introducing pitch changes. Connected paragraphs accelerated in this way remain highly intelligible at rates up to 400 words per minute (about 20 to 30 ms per phoneme), and intelligibility remains as high as 50% when monosyllabic words are reduced in duration by 75% to 85%. By way of contrast, *producing* speech at comparable rates and levels of intelligibility is well beyond human capacity.

The usual explanation of why perception is so much faster than production is that extra time and effort are required to physically move articulators such as the jaw. However, this explanation of the maximal rate asymmetry conflicts with my 1981 data on the rate of internal speech (D. G. MacKay, 1981). I had subjects produce sentences internally (silently without moving the lips) as rapidly as possible, pressing one key when they began a sentence and another key when they ended it. The maximal rate, measured at asymptote after many practice trials with the same sentence, was about 91 ms per phoneme. This rate is much faster than the maximal rate of overt speech. Subjects producing identical sentences *aloud* as quickly as possible only achieve an asymptotic rate of about 133 ms per phoneme after many trials of practice. Nevertheless, although faster than overt speech, the maximal rate of internal speech is considerably slower than the 20- to 30-ms per phoneme rate during perception of time-compressed speech. Because the articulators do not move during internal speech, this remaining rate difference indicates that muscle movement factors cannot completely explain the maximal rate asymmetry. A satisfactory account requires reference to fundamental differences between processes underlying perception versus production.

The Listening Practice Asymmetry

D. G. MacKay and Bowman (1969) reported a "conceptual practice effect," which has an interesting but asymmetric counterpart on the perceptual side, reported here for the first time. The subjects were German–English bilinguals, who were presented with sentences one at a time and simply produced each sentence as rapidly as possible. An example is "In one corner of the room stood three young men." Following a 20-s pause, the same sentence was presented again, for a total of 12 repetitions or practice trials with each sentence. There were 12 different sentences in all, and for reasons that will become clear shortly, six were in English and six in German. The independent variable was trial of practice, and the dependent variable was the time to produce the sentence.

Results for this "physical practice condition" appear in Table 6.2 and are averaged over the first four and last four practice trials for 12 sentences. As can be seen in Table 6.2, speech rate was 15% faster for the last four than for the first four practice trials, even though the subjects were attempting always to speak at their maximum rate.

Twenty seconds after the twelfth repetition of a "practice" sentence, the subjects received a "transfer" sentence in their other language, which they also produced at maximal rate for four trials. This transfer sentence was either a translation or a nontranslation of the practice sentence. Nontranslations were unrelated to the original sentence in meaning, syntax, and phonology, while translations had the same meaning as the original but differed in phonology and word order. To control for sentence difficulty, the transfer sentences were counterbalanced across subjects, so that exactly the same sentence occurred as either a translation or a nontranslation, depending on the nature of the practiced sentence.

The subjects produced each transfer sentence four times, again with 20 s between repetitions, and these data are averaged across repetitions for the 12 transfer sentences in Table 6.2. The rate of speech for translations was 8% faster than for nontranslations (2.44 s per sentence versus 2.24 s per sentence), a statistically reliable difference indicating an effect of practice at the conceptual level in speech production (see D. G. MacKay, 1982, for a detailed explanation).

Consider now the perceptual analogue of this conceptual practice effect. The critical condition involved *listening practice*, which was designed to determine whether repeated *listening* to a sentence leads to facilitation in the same way as repeated *production*. Twelve German–English bilinguals listened to a tape recording of the subjects discussed previously who had produced the sentences

TABLE 6.2. Practice and transfer effects (in secs per sentence) for the physical practice and listening practice conditions.

	Production times (per sentence)					
	Practice condition		Transfer condition			
	First 4 trials	Last 4 trials	Non-translation	Translation	Difference	% Facilitation
Physical practice	2.33	2.03	2.44	2.24	.20	8%
Listening practice	2.33	2.03	2.57	2.31	.26	10 %

12 times at maximal rate. To ensure that "listening practice" subjects were paying attention to the sentences in this condition, the listeners were instructed to indicate whether "physical practice" speakers made changes or errors from one repetition to the next. (No such errors actually occurred.)

A transfer phase, identical to that in the physical practice condition, followed each set of 12 listening practice trials. In this transfer phase of the listening practice condition, subjects produced out loud and at maximal rate a sentence that was either a translation or a nontranslation of the sentence that they had heard repeated 12 times. The results appear in Table 6.2. As before, production times were faster for translations (2.31 s), than for nontranslations (2.57 s) in the transfer phase of the listening practice condition, a 10% facilitation effect compared to the 8% facilitation effect for the physical practice condition. However, there was an asymmetry in the *absolute* production times for transfer sentences in the physical versus listening practice conditions. Absolute production times (see Table 6.2) were significantly longer in the listening practice condition than in the physical practice condition, both for nontranslations (6% longer) and for translations (3% longer).

Errors in Perception Versus Production

Even though production has received much less attention than perception in the field at large (Chapter 1), slips of the tongue have been collected and studied much more often than slips of the ear, the perceptual errors that occur when listening to conversational speech (Fromkin, 1980). However, studies of misperceptions and misproductions have been undertaken for the same basic reason. Regularities in misperceptions and misproductions allow inferences about the otherwise hidden mechanisms underlying everyday perception and production, and they "test" existing theories, because theories that are incapable of explaining the errors that occur are incomplete or inadequate as accounts of mechanisms underlying "veridical" perception and production (see, e.g., Bond & Garnes, 1980; Freud, 1901/1914; Meringer & Mayer, 1895). Interestingly, however, systematic comparisons of misperceptions and slips of the tongue do not currently exist. What follows are some preliminary comparisons suggesting six general classes of asymmetries for further empirical test: sensory asymmetries, muscle movement asymmetries, phonological asymmetries, sequential asymmetries, lexical asymmetries, and semantic asymmetries.

Sensory Asymmetries

Sensory factors contribute to perceptual errors but have no analogous effect on production errors. An example is the masking asymmetry. Extraneous environmental noises often cause misperceptions by masking incoming speech sounds at the sensory analysis level, but speech errors directly attributable to extraneous (nonspeech) sounds have never been reported.

Muscle Movement Asymmetries

Whole classes of production errors lack a perceptual analogue. An example is stuttering, a class of speech errors that simply never occurs in perception. Listeners never misperceive someone to say *p-p-p-please* when the speaker in fact said *please*. This asymmetry is one of many sources of converging evidence suggesting that "intrinsic" stuttering (D. G. MacKay & MacDonald, 1984; and Chapter 8) originates within the muscle movement system.

Phonological Asymmetries

The Syllabic Position Asymmetry

Phonological production errors obey a syllabic position constraint (Chapter 3). Syllable-initial segments substitute with other syllable-initial segments and never with syllable-final or -medial segments. However, perceptual metatheses, for example, *know* misperceived as *own* (Bond & Garnes, 1980), are reversals in the order of segments within a syllable that violate this syllabic position constraint. Similar production errors involving transposition of a vowel and a consonant have never been collected from *adult* speech.

The Phonological Similarity Asymmetry

The phonological similarity asymmetry refers to the greater role of phonological similarity in misperceptions than in misproductions. Misproductions *sometimes* involve phonologically similar words such as *carcinoma* and *Barcelona* (Dell, 1980), but misperceptions virtually always involve phonologically similar words (Browman, 1980).

Sequential Asymmetries

The Sequential Domain Asymmetry

The sequential domain asymmetry refers to the greater role of syntactic class in misproductions than misperceptions. Misproduced words almost invariably substitute words from within the same syntactic class, for example, nouns interchange with other nouns and virtually never with verbs or adjectives (Chapter 3; Cohen, 1967; D. G. MacKay, 1979), but misperceptions frequently violate this syntactic class constraint. An example violation is the misperception of *descriptive* as *the script of*, where a determiner, a noun, and a preposition substitute an adjective (example from Bond & Garnes, 1980).

The Sequential Error Asymmetry

Sequential errors are the most common class of production error (Chapter 2) and include anticipations, preservations, and transpositions of about-to-be-uttered words and speech sounds. Two of Fromkin's (1973) examples are the phonological transposition *coat thrutting* for *throat cutting* and the word reversal "*We have a laboratory in our computer*" for "*We have a computer in our laboratory.*" The sequential error asymmetry refers to the fact that perceptual errors resembling these high-frequency production errors have never been reported.

Lexical Asymmetries

The Word Substitution Asymmetry

The word substitution asymmetry refers to the greater frequency of word-for-word substitution errors in perception than in production. Units involved in misperceptions range in scope from entire phrases (e.g., *popping really slow* misperceived as *prodigal son*), to single features (e.g., *pit* misperceived as *bit*), to word substitutions, in which the listener mishears one word as another. These word-for-word substitutions predominate over other errors in collections of misperceptions (85% versus 15% in Browman, 1980), whereas the opposite is true in collections of speech errors. For example, word-for-word substitutions made up only 38% of the Garnham, Shillcock, Mill, and Cutler (1982) corpus, which includes every speech error from a large sample of recorded speech and is especially suited for this type of comparison.

The Lexical Error Asymmetry

The lexical error asymmetry refers to the fact that nonwords appear more often as misproductions than as misperceptions. Listeners almost invariably misperceive words as other words (Browman, 1980), whereas speakers often misproduce words as nonwords. For example, 63% of the phonological production errors in Dell's (1980) corpus resulted in nonwords, such as *thrutting* instead of *cutting*.

Position-Within-A-Word Asymmetry

The position-within-a-word asymmetry refers to the tendency for misproductions and misperceptions to involve different parts of a word. That is, in both perception and production, some parts of a word are more susceptible to errors than others, but the parts susceptible to perceptual errors differ from the parts susceptible to production errors. Word-initial segments are more likely to be misproduced than word-middle segments (D. G. MacKay, 1970e), whereas word-middle segments are more likely to be misheard than word-initial segments (Browman, 1980).

Word Boundary Asymmetry

In word boundary errors the juncture between words is mislocated, as in the misperceptions *tenure* for *ten year*, and *take a pillow* for *take a pill out*. The word boundary asymmetry refers to the fact that these errors are relatively common in perception, making up about 18% of the 1980 Garnes and Bond corpus, but they simply do not appear in collections of tongue slips. Production errors resembling "*They had a ten-year party. . . . Excuse me, I mean, a tenure party for Marlene*" have never been reported.

Semantic Asymmetries

The Garden Path Asymmetry

The garden path asymmetry refers to the absence of garden path errors in production. Garden path miscomprehensions (Chapters 1 and 4) are relatively common. Listeners often perceive the wrong meaning of an ambiguous word, such as, say, *crane*. However, speech errors resembling garden path miscomprehensions have never been reported. Normal speakers never begin to discuss one type of *crane*, say, *machine cranes*, and then *inadvertently* end up discussing *bird cranes*.

The Synonymic Asymmetry

The synonymic asymmetry refers to the absence of synonymic errors or blends in perception. A typical synonymic error is "*He was sotally responsible for that*, a blend of "*He was solely responsible for that*" and "*He was totally responsible for that*." Blends are relatively common among speech errors, making up 15% of all word errors in Garnham et al. (1982), but blends lack a perceptual analogue. For example, listeners never mishear *solely* as a combination of *solely* and *totally*.

Methodological Issues

Error collections pose well-known analytic difficulties for both misproductions (D. G. MacKay, 1980) and misperceptions (Bond & Small, 1984; Cutler, 1982), and comparisons between misperceptions and misproductions can only compound these difficulties. Listener–collectors can't observe and record misperceptions *directly*. Misperceptions must be inferred in one way or another, often from pragmatic context, and it seems likely that many more misperceptions than misproductions go undetected by researchers and lay people alike (all other factors being equal; Warren, 1982). Moreover, some production errors may be more difficult to perceive (and collect) than others (Bond & Small, 1984; Cutler, 1982). This means that collections of misproductions and misperceptions may be nonindependent and statistically incomparable on a priori grounds and that the asymmetries discussed previously require further empirical support. What is needed are laboratory techniques for inducing misperceptions and misproductions experimentally. A prototype technique of this sort has already been developed (Chapter 2; and D. G. MacKay, 1978), but so far has only been applied to a single species of phonological errors. Extending such techniques to a wider range of experimentally induced misperceptions and misproductions could both test and extend the list of asymmetries discussed previously.

Structural Asymmetries in the Node Structure Theory

I turn now to asymmetries in the node structure theory which explain the empirical asymmetries discussed previously. I begin with structural asymmetries: Top-down connections differ from bottom-up connections in four fundamental ways in the node structure theory. One of these theoretical asymmetries reflects a difference in the distribution of top-down versus bottom-up connections in the network at large, the **uniqueness asymmetry**; another reflects a difference in the relative strength of top-down versus bottom-up connections, the **linkage strength asymmetry**; and two reflect the independent status of sensory analysis versus muscle movement nodes in the theory, the **sensory analysis and muscle movement asymmetries**.

The Muscle Movement Asymmetry

The fact that muscle movement nodes play a role in production but not perception contributes directly to the stuttering asymmetry (Table 6.1). As discussed in Chapter 8, intrinsic stuttering is a disturbance that causes errors in the muscle movement system but nowhere else.

The Sensory Analysis Asymmetry

The fact that sensory analysis nodes play a role in perception but not production is directly responsible for the masking asymmetry (Table 6.1). Extraneous environmental noises introduce a disturbance within the sensory analysis system that causes misperceptions but not misproductions.

The Uniqueness Asymmetry

The **uniqueness asymmetry** refers to a difference in the distribution of top-down versus bottom-up connections to domains in the node structure network. Top-down connections to content nodes in a domain are generally unique or singular, so that only a single node in a domain normally receives top-down priming at any given time during production. On the other hand, bottom-up connections to content nodes are massively nonunique. Any given input simultaneously transmits bottom-up priming to many different nodes in many different domains. This **uniqueness asymmetry** contributes to at least three empirical asymmetries: The sequential domain asymmetry, the word boundary asymmetry, and the garden path asymmetry.

Sequential Domain and Word Boundary Asymmetries

Production errors rarely violate the sequential class constraint, because each lexical node receives priming from a unique source and passes on this unique priming to a single sequence node that determines the syntactic class of what gets produced. Errors can only occur when an extraneous node within the same sequential domain as the intended word achieves most priming when the sequence node becomes activated. As a result, words substitute in error with words from the same sequential domain or syntactic class, which is the syntactic class constraint.

However, bottom-up connections do not confine *perceptual* alternatives to a single sequential domain in the theory. Because a given input can transmit bottom-up priming nonuniquely to nodes in many different domains, an extraneous sequence node can receive greatest priming and become activated in violation of the syntactic class constraint. For example, when perceiving the word *descriptive*, *descriptive*(adjective) may receive and pass on less priming to its sequence node than does *the*(determiner), *script*(noun), and *of*(preposition), in part because of the lexical frequency of *the* and *of*, but perhaps also because of top-down (expectation) priming of *script*(noun). As a consequence, DETERMINER, NOUN, and PREPOSITION will become activated as the most primed sequence nodes, rather than ADJECTIVE, a multiple violation of the syntactic class constraint. As this same example illustrates, **the uniqueness asymmetry** also contributes to the word boundary asymmetry, the fact that word boundaries are subject to error in perception but not production.

The Garden Path Asymmetry

The nonuniqueness of bottom-up connections also contributes to the garden path asymmetry (see also Chapter 7). A lexically ambiguous word such as *crane* causes garden path miscomprehensions (D. G. MacKay, 1970d) because its syllable node, *crane*(stressed syllable), connects with and primes two lexical content nodes representing machine cranes and bird cranes, so that the wrong node can become activated under the most-primed-wins principle. However, the uniqueness of top-down connections prevents similar errors in production. Top-down connections from a node such as *the tall crane*(noun phrase) go to either *crane 1*(noun) or *crane 2*(noun) but not both. As a result, speakers can't intend to discuss bird cranes and *inadvertently* end up discussing machine cranes, because only the lexical node for machine cranes or for bird cranes receives top-down priming.

The Linkage Strength Asymmetry

The **linkage strength asymmetry** refers to the fact that bottom-up connections tend to be stronger than top-down connections, especially for higher level nodes, because we generally perceive words more often than we produce them: Listening–reading is a more common activity than speaking–writing–typing. Together with several other theoretical asymmetries, this **linkage strength asymmetry** contributes to the maximal rate asymmetry, which is the fact that perception can proceed faster than production, and to the word production asymmetry, which is the fact that we comprehend words long before we use them in speech production.

Processing Asymmetries in the Node Structure Theory

Although identical mental nodes and microprocesses play a role in perception and production, macroprocesses for perception and production exhibit four fundamental asymmetries, summarized here as the **priming summation asymmetry**, the **level of activation asymmetry**, the **sequential activation asymmetry**, and the **connection formation asymmetry**.

The Priming Summation Asymmetry

The **priming summation asymmetry** arises from the fact that top-down priming *in action hierarchies* diverges via one-to-many connections, whereas bottom-up priming *in perceptual hierarchies* converges via many-to-one connections. This **priming summation asymmetry** contributes to at least six empirical asymmetries: the phonological similarity asymmetry, the maximal rate asymmetry, the word boundary asymmetry, the synonymic asymmetry, the position-within-a-word asymmetry, and the sequential error asymmetry.

The Phonological Similarity Asymmetry

The **priming summation asymmetry** ensures that misperceptions involve phonologically similar words more often than misproductions do. Bottom-up priming converges and summates to such an extent on the input side that misperceptions must incorporate most of the phonological components of the actual input. However, on the output side, bottom-up priming only converges on just-activated nodes, which are undergoing self-inhibition. As a result, only *divergent* bottom-up priming can introduce phonological similarity into misproductions, and only rarely because divergent priming is second order and relatively weak.

The Maximal Rate Asymmetry

The **priming summation asymmetry** is another contributor to the maximal rate asymmetry. Because priming converges and summates to a greater extent in perception than in production, mental nodes can be activated more quickly in perception than in production (for a given error criterion; D. G. MacKay, 1982).

Sequential Error Asymmetry

The **priming summation asymmetry** also explains why sequential errors are much more common in production than in perception. The anticipatory priming that results from divergent top-down connections readies upcoming units for activation and increases the probability of anticipatory errors in production. Now, anticipatory priming also occurs occasionally in perception, and misperceptions sometimes reflect what word the listener expects a speaker to say (Garnes & Bond, 1980). The difference is that anticipatory priming invariably occurs at all levels of production, but only sometimes, and only at the word level during perception. That is, perceptual expectations center on word concepts and not on units at higher or lower levels. We normally cannot anticipate the phonemes in upcoming words during perception, a necessary condition for sequential misperceptions resembling production errors such as *coat thrutting* for *throat cutting*. Nor can we generally anticipate phrases in perception, a necessary condition for sequential misperceptions resembling speech errors such as "laboratory in our computer" for "computer in our laboratory."

The **priming summation asymmetry** also promotes the sequential error asymmetry in another way. The fact that convergent summation is everywhere present in perception, but not in production, strongly constrains perceptual errors to resemble the actual phonological input, thereby initially ruling out perceptual substitutions of phonologically dissimilar words as in the preceding *laboratory–computer* example.

Position-Within-A-Word Asymmetry

The **priming summation asymmetry** also contributes to the position-within-a-word asymmetry, the fact that word-initial segments are more likely to be misproduced than word-middle segments, whereas word-middle segments are more likely to be misheard than word-initial segments. As already noted, the perceptual effect reflects the "left-to-right" summation of convergent priming which occurs in perception but not production, and the production effect reflects summation of divergent (anticipatory) priming which occurs in production but not perception.

The Synonymic Asymmetry

The **priming summation asymmetry** also contributes to the synonymic asymmetry, the fact that blends occur in production but not perception. Blends result under the theory whenever context enables two or more nodes in the same domain to receive exactly equivalent priming at the time when the activating mechanism is applied, so that both nodes become activated simultaneously (D. G. MacKay, 1973b; and Chapter 2). For example, if both *solely*(adverb) and *totally*(adverb) receive equivalent priming when ADVERB is activated in producing the sentence "He was solely/totally responsible for that," an error such as *sotally* will occur. Whatever phonological nodes receive more priming from either *solely*(adverb) or *totally*(adverb) or both will become activated. However, blends such as *sotally* cannot occur in perception, because bottom-up priming will converge on either *solely*(adverb) or *totally*(adverb) but not both.

The Level of Activation Asymmetry

In everyday production, both higher and lower level nodes must become activated because the output must be produced in sequence. In everyday perception, however, only higher level nodes must become activated. Because of the strength of lower level connections and the convergent summation and overlapping timing characteristics of lower level bottom-up priming, lower level nodes need not become activated in order to pass on sufficient priming to reach commitment threshold of their connected nodes. As noted in Chapter 4, this principle of higher level activation is flexible in its application, and achieving this flexibility is one of the reasons why nodes are organized into modalities, systems, and domains (see Chapters 2 and 5).

The Maximal Rate Asymmetry

The **level of activation asymmetry** is a major contributor to the maximal rate asymmetry. Because all nodes become activated in production, whereas only higher level nodes normally become activated in perception, perception can proceed faster than production in the theory.

Lexical Error Asymmetry

The **level of activation asymmetry** also helps explain the lexical error asymmetry, the fact that nonwords result when speakers but not listeners make phonological errors. Because phonological units don't normally become activated in everyday speech perception, phonological errors resulting in nonwords are rare in perception.

The Sequential Activation Asymmetry

Order of activation is asymmetric under the node structure theory. Even when the same higher level nodes become activated during both perception and production, order of activation differs in perception versus production. As noted in Chapter 2, mental nodes that become activated in the order 1, 2, 3, 4, 5 during production will become activated in the order 3, 4, 2, 5, 1 during perception in the node structure theory (see Figure 2.2). Under Bierwisch's (1966/1985) symmetry assumption, however, production and perception might be expected to exhibit reverse orders of activation, that is, 1, 2, 3, 4, 5 during production, with the extremely unlikely reverse order, 5, 4, 3, 2, 1, during perception (see Figure 2.2).

The Connection Formation Asymmetry

A final asymmetry between bottom-up and top-down processes is that new connections between nodes are initially formed bottom-up rather than top-down in the node structure theory. As already noted, I discuss the process of connection formation elsewhere (D. G. MacKay, 1987), rather than in the present book. However, I mention the **connection formation asymmetry** here not just for the sake of completeness but for its potential role in the listening practice and word production asymmetries. The main reason that listening practice facilitates performance (see Table 6.2) may be because the relevant content nodes and their bottom-up connections become formed and strengthened during listening practice. However, listening practice facilitates performance *less* than physical practice (the listening practice asymmetry), because top-down connections do not become formed, activated, and strengthened during listening practice. Similarly, children can recognize and understand a word long before they can produce it, because activating a lexical content node with bottom-up connections from phonological and sensory analysis nodes suffices for recognizing a word under the theory. However, *producing* the word requires formation of several additional types of connections, each of which may delay development of the production vocabulary. One additional type of connection is top-down from the lexical content node to the appropriate phonological and muscle movement nodes. Another is the inhibitory connection between sequence nodes required for sequencing the phonological and muscle movement components for producing the word.

The Aphasic Asymmetry Prediction

Asymmetries in the node structure theory generate a large number of predictions. I discuss the aphasic asymmetry prediction as one example. Under the theory, production and perception will break down symmetrically in most aphasias. Most lesions will damage content nodes, causing symmetric impairment to production and comprehension under the theory, because content nodes are essential to both. Even when sequence and timing nodes are selectively damaged, leaving content nodes intact, effects will sometimes be symmetric. For example, selective damage to *sentential* sequence and/or timing nodes will impair production and comprehension of sentential components symmetrically, and will leave *phonological* perception and production intact: The patient will be able to produce and recognize phonological components without comprehending their meaning and will be able to identify and repeat acoustically presented nonwords.

However, selective damage to *phonological* sequence and/or timing nodes (again leaving content nodes intact) will only impair *production*, and in a strikingly distinctive way. The reason for this aphasic asymmetry prediction is that phonological sequence and timing nodes are unnecessary for perceiving common words because phonological nodes become activated during production but not during perception (the **level of activation asymmetry**). Moreover, when this aphasic asymmetry occurs, the theory predicts a severe and unique type of production deficit. Activation, sequencing, and/or timing of phonological components will be impaired, with unusual intonation and rhythm at the phonological level, and large numbers of spoonerisms and segment distortions, but no corresponding perceptual errors.

7
The Functions of Mental Nodes

> The ultimate function of all neural analysers of sensory input is not mere description or classification, but the shaping of conditional readiness to reckon with the state of affairs betokened by that input. The main question to be answered by the sensory system is not 'What is it?' but 'What does it signify for me?,' or if you like, 'So what?'
>
> (D. G. MacKay, 1984, p. 262)

> The message received may be subject to perceptual errors and confusions resulting from environmental noises, unfamiliarity with the speaker's pronunciation and style, false expectations of the speaker's intent, unintentional ambiguities in what the speaker is saying. . . .
>
> (Warren, 1982, p. 177)

Previous chapters have advanced various sources of evidence for the existence of nodes that play a role in both perception and action. Chapter 7 examines the functional issues. Why have mental nodes evolved? What functions do mental nodes serve?

I examine two possible answers to these questions. One concerns structural economy. Mental nodes are multipurpose processors, and using nodes for more than one purpose may serve to economize on the number of nodes. As we will see, this answer turns out to be surprisingly weak. The other answer concerns the integration of different types of information and turns out to be much stronger. I argue that mental nodes have evolved to enable the rapid integration of heterogeneous sources of information, not just across perception and action but across a variety of other sensory and conceptual modalities. I then examine some of the costs and benefits of this rapid integration process, for example, errors in perception and action (cost) and the automatic resolution of ambiguity (benefit).

The Structural Economy Hypothesis

Two types of economy must be distinguished when addressing the economy issue: structural economy and processing economy (see also Collins & Quillian,

1969; Grossberg, 1982). Structural economy refers to the number of nodes that play a role in some activity, whereas processing economy refers to the number of processing operations that these nodes participate in and how fast these processing operations can be carried out.

Under the structural economy hypothesis, using the same nodes for more than one purpose economizes on structure. Fewer nodes and connections between nodes are required to carry out any given function. In the case of mental nodes, the structural economy hypothesis would be true by definition if perceptual processes were simply the reverse of the corresponding production processes and if no additional mechanisms were required to prevent interactions between top-down and bottom-up processes involving identical nodes.

Unfortunately, both of these prerequisites to easy acceptance of the structural economy hypothesis are false. Chapter 6 presented theoretical and empirical arguments showing that perceptual processes are not simply the reverse of the corresponding production processes, and Chapters 8 through 10 will present convincing evidence that at least one additional mechanism (self-inhibition) is necessary for preventing unwanted interactions between top-down and bottom-up processes involving mental nodes. Without self-inhibition during production, bottom-up priming can cause inadvertent reactivation of higher level mental nodes. Moreover, all but the lowest level perception–action nodes require this self-inhibitory mechanism, a fact that makes the structural economy argument difficult to evaluate and sustain. Mental nodes incur a hidden structural cost and must have evolved for other reasons besides structural economy.

Processing Economy: Integration of Heterogeneous Information

The other type of economy is processing economy. Mental nodes may economize on the rate of processing or on the processing steps required for perception and action. In what follows, I argue that mental nodes automatically integrate heterogeneous sources of information, not just for perception and action but for different sensory modalities such as vision and audition and for different sources of information originating within more cognitive systems (see also Morton, 1969). I argue that, by economizing on processing, and by automatically resolving perceptual ambiguities, broadly defined, this automatic integration process has contributed to the evolution of mental nodes.

The Integration of Perception and Action

That perception becomes integrated with action is self-evident, because perception carries out monitoring functions that are required for the regulation of action (Chapter 9). How and to what extent perception becomes integrated with action is not self-evident, however, and represents a central topic of this book. In the present chapter, I show how mental nodes enable a complete and rapid mesh between perception and action in general, and in Chapter 9, I show how mental nodes facilitate the processing of perceptual feedback during ongoing action.

In the case of mental nodes, perception is synonymous with a disposition to respond. When a mental node becomes activated during perception, all of its associated higher level (e.g., proposition) nodes and lower level (e.g., phonological) nodes become strongly primed or readied for activation under the most-primed-wins principle. And because priming is necessary for action, activating a mental node in perception can be considered the first stage in preparation of a response. Mental nodes activated during perception prime a wide range of possible responses via previously formed connections, both bottom-up and top-down. The top-down priming to lower level nodes enables a repetition response, as occurs during shadowing (Chapter 2), and the bottom-up priming to higher level nodes enables more complex, propositional responses. Following node activation during perception, whatever higher level nodes have accumulated most priming from other internal and external sources can be quickly and automatically activated under the most-primed-wins principle to generate the contextually most appropriate response.

Integration Across Sensory Modalities

Besides integrating perception and action, mental nodes integrate information across sensory modalities. Events in nature usually stimulate more than one sensory system, and mental nodes enable the rapid integration of this correlated information so as to optimize activities within the environment. A typical example is the integration of optical cues to depth with proprioceptive cues originating in the muscles responsible for changing the vergence of the eyes and altering the focal curvature of the lens (Steinbach, 1985). As Warren (1982) points out, these proprioceptive cues "are integrated with, and are indistinguishable from, the purely optical cues to depth. For a person perceiving an object at a particular distance, these (proprioceptive) cues are as fully visual as those providing information via the optic nerve" (p. 189). Interactions between the sight and sound of moving lips during speech perception (the McGurk effect discussed below and in Chapter 2) illustrate another functionally useful integration of information originating in different sensory modalities.

Visual Dominance in Cross-Modality Integration

Mental nodes not only combine information from different senses, they determine the perceptual outcome that results when different sources of sensory information conflict. Mental nodes therefore form an integral part of the explanation of why visual information generally dominates, and determines the nature of the resulting experience, when visual inputs conflict with inputs from any other sensory system. I explore the possible role of mental nodes in three types of visual dominance effect: visual dominance in maintaining balance, localizing inputs, and perceiving speech.

Visual Dominance Over Vestibular Information

Vision dominates in cross-modality integrations with vestibular information (Graybiel, Kerr, & Bartley, 1948; D. N. Lee & Lishman, 1975), and the unanswered question is why. An interesting possibility under the node structure theory focuses on the relative number of low-level nodes devoted to vision versus vestibular sensations. Because retinal neurons greatly excede vestibular neurons in number (Mittelstaedt, 1985), it can be argued that mental nodes receive greater convergent priming from visual inputs than from corresponding vestibular inputs. As a result, when visual and vestibular information conflicts, visual input will contribute more priming, and thereby determine which mental nodes become activated and give rise to perception. Dominance effects between different sorts of information originating within a sensory modality—for example, sacular versus utricular information within the vestibular system—can likewise be explained in terms of relative number of converging neurons (Mittelstaedt, 1985). However, this "spatial convergence" argument can only be applied to low-level, highly automatic perceptual processes. Because the principle of higher level activation is flexible (within limits, Chapter 4; and D. G. MacKay, 1987), the nervous system can readily modify its weighting of different kinds of evidence at *higher levels* (see also Grossberg, 1982).

Visual Dominance in Spatial Localization

Vision also dominates over other sensory modalities in perceiving the spatial location of inputs. For example, when corresponding visual and auditory inputs arrive from conflicting spatial locations, vision usually determines perceived localization. An example is the "ventriloquism effect." Sounds from a concealed loudspeaker can be displaced up to 20 degrees from a visible sound source (e.g., an actor's moving lips), but subjects continue to perceive the sounds as coming directly from the visual source (Witken, Wapner, & Leventhal, 1952). This and many other visual dominance effects are usually explained in terms of the relative reliability of vision versus audition. Under this hypothesis, we tend to rely more on vision than audition, because vision is more accurate than audition for spatial localization. However, some interesting exceptions to visual dominance suggest that this explanation is incomplete or inadequate. For example, neither vision nor audition dominates when the sound track of a film is out of synch or poorly dubbed. We both perceive and are bothered by the asynchrony between visual and auditory events (Neisser, 1976). If people can simply tune audition out during spatial localization, why can't they tune audition out in the case of the poorly synchronized sound track?

The node structure theory provides a more complete explanation of both visual dominance and its exceptions. Phrased within the present framework, the basic issue is this: Why do visual connections to mental nodes generally contribute more priming than auditory or tactile connections? I will discuss two possible answers to this question. One is that for some types of inputs, visual connections have received greater prior practice than connections from other sensory systems (see the discussion of visual skill in D. G. MacKay, 1987). The other is that mental nodes receive greater convergent priming from vision than from any other sensory system. There are simply more nodes in the visual system that can contribute priming, and the convergent priming from these visual nodes is usually simultaneous and continuous in nature, unlike the priming from, say, auditory nodes, which is usually sequential and discontinuous. As a result, visual nodes contribute more spatially and temporally summated priming than do auditory inputs, thereby dominating perception under the most-primed-wins principle.

These hypotheses predict that exceptions to visual dominance will arise when visual and auditory sources have received comparable or asymptotic degrees of prior practice, are comparably discontinuous or sequential in nature, and produce convergent priming from the same number of nodes. Such is surely the case for the bothersome asynchrony between auditory speech and visual lip movements for a poorly synchronized sound track. Both the auditory and the visual inputs arising from the moving lips are discontinuous and sequential, and the perception of timing for visual and auditory speech events probably engages identical timing nodes (Chapter 5). In addition, prior effects of practice for both visual and auditory representations of speech sounds are undoubtedly asymptotic (D. G. MacKay, 1982). The absolute amount of practice for visual and acoustic speech events is so great that the added practice that audition receives through use of, say, telephones and radios provides no additional benefit. As a result, mental nodes will receive nearly equivalent priming from visual and auditory representations of speech sounds, so that neither vision nor audition can dominate under the most-primed-wins principle. The out-of-synch sources will engage in continuous and bothersome conflict with no possibility of automatic resolution.

The fact that timing nodes for vision and audition are shared also predicts that visual dominance will evaporate in a task where subjects estimate the duration of visual and auditory signals, presented either separately or together. Under the node structure theory, dominance depends on the detailed nature of the task being performed and not on fixed dominance relations between different input systems or on general attentional strategies (as in Posner, 1978).

Visual Dominance in Speech Perception

When observing someone speaking, vision dominates when what we hear and what we see conflict in content (rather than in timing). For example, when subjects watch a video recording of a speaker producing a visually distinctive syllable such as *pa* while hearing the syllable *ta* dubbed in synchrony onto the sound track, they usually report hearing *pa* (McGurk & MacDonald, 1976). This visual effect is both unconscious and automatic. Subjects are unaware that the *pa* they "*heard*" originates in the visual signal, and they are unswayed by either instructions or personal experience to the contrary. That is, the illusion persists even when subjects are informed that the auditory and visual inputs differ, and even when they close and open their eyes and observe (to their surprise) that their perception alternates between the visual alternative (with the eyes open) and the auditory alternative (with the eyes closed).

Why does vision dominate in this case, but not in the case of the out-of-synch sound track? As discussed, the fact that we see and hear speech sounds produced with asymptotic levels of practice rules out an explanation in terms of the relative frequency of visual versus auditory "speech events." The fact that both the auditory and the visual inputs arising from the moving lips are discontinuous and sequential also rules out an explanation in terms of temporal convergence of priming. This leaves the spatial convergence hypothesis (discussed previously for vestibular inputs), in which visual nodes outnumber acoustic analysis nodes, so that when the two content sources conflict, visual inputs contribute more convergent priming than auditory inputs and thereby determine which mental nodes become activated and give rise to perception. That is, when a segment is visually distinctive as in McGurk and MacDonald (1976), nodes representing effects of lip movements within sensory analysis systems for vision and audition both converge bottom-up on a set of phonological nodes; but because of their greater numbers, visual nodes contribute more priming, and determine which phonological node receives most priming and becomes activated under the most-primed-wins principle.

Integration Within a Sensory Modality

Mental nodes also integrate different sources of information originating *within* a sensory–cognitive module. For example, the Stroop effect illustrates how mental nodes integrate two sources of visual information: color and orthographic form. Subjects in the Stroop task must name the color of ink that a color word is printed in, and their reaction times are especially fast when color (e.g., "red") and word (*red*) are identical (Keele, 1973). The reason is that both inputs prime the same lexical content node, *red*(color adjective), which is therefore readily activated under the most-primed-wins principle. However, errors and reaction times increase dramatically when the color and the color word differ, because the two sources of visual input conflict and prime different mental nodes in the same (color adjective) domain, the prototypical situation for the occurrence of errors under the theory.

Integration Across Cognitive Modalities

Language is an integrative module par excellence, and can be seen to join cognitive modalities as well as sensory and motor modalities per se. When we say that we like what we see, hear, smell, taste, feel, or any combination of these, language becomes a common source of conceptual integration that spans these sensory–conceptual sources. The way that the smell or sight of apples virtually immediately can evoke the name *apple* also illustrates how language integrates concepts from different sensory–conceptual sources. Experiments on ambiguous figures further illustrate how quickly speech and visual concepts can become integrated. Presenting a word such as *duck* has extremely rapid effects on the visual interpretation of an ambiguous figure such as Jastrow's rabbit-duck (Leeper, 1936).

Mental nodes are the basis for these rapid cross-modal integrations and also enable efficient integration of concepts within the language module. When interpreters translate on-line between languages, for example, they rapidly integrate one set of language concepts with another. A similar integration process occurs at a lower level during shadowing, in which mental nodes rapidly integrate acoustic speech sounds with corresponding muscle movements for producing them. Studies of expert shadowing (e.g., Marslen-Wilson, 1975) demonstrate how rapidly this particular sensory–motor integration can occur. Trained shadowers can repeat back words in sentences with lags of less than 300 ms and begin to reproduce a long word after hearing only its first few segments. Top-down semantic and pragmatic information can also enter into the integrations that occur during "close shadowing." Close shadowers "repair" intentional mispronunciations of the first segment of an input word, reconstructing the original word top-down on the basis of semantic and pragmatic context (Cole & Scott, 1974). These short-lag repairs suggest that semantic and pragmatic processing has an extremely rapid, on-line effect during word recognition, and mental nodes provide a mechanism for explaining how such high-level information can be brought to bear so quickly.

McLeod and Posner (1981) argued that the auditory–vocal integration that occurs in shadowing reflects a "privileged loop" that makes phonology special. In their experiments, shadowing the phonology of a word such as *high* (stimulus)–*high* (response), enabled interference-free dual task performance, whereas producing semantic associates such as *high–low* did not. Under the node structure theory, mental nodes are the mechanisms underlying privileged loops, and it should be possible to observe similarly privileged loops in other skills, levels of language, and aspects of skill. For example, a formally identical integration process occurs at sentential levels during skilled on-line translation between languages. Shared mental nodes provide a privileged loop that enables rapid integration of language 1 input with language 2 output at the sentential level (D. G. MacKay, 1981; 1982). Timing nodes likewise form part of a privileged loop, enabling interference-free execution of many concurrent activities, as when we march, sing, and breathe in time with the band (Chapter 5). Other skills such as expert transcription typing and expert Morse code also involve mental nodes (D. G. MacKay, 1985) that form the basis for other privileged input–output loops under the node structure theory. Indeed, privileged loops (mental nodes) may provide the underlying basis for all highly compatible input–output relations, such as a finger-press response to tactile stimulation of the same finger.

Automatic Resolution of Ambiguity: A Benefit of Integration

To summarize the chapter so far, mental nodes rapidly integrate many different types of information, with perception and action being only the most notable examples, and this integrative process may have contributed to the evolution of mental nodes. However, the automatic integration of information via mental nodes has two inadvertent side effects. I have already mentioned one (visual dominance), and I discuss the other (errors) in the next section. Here I examine how mental nodes disambiguate inputs by rapidly and automatically integrating huge amounts of heterogeneous contextual information. I will argue that contextual resolution of ambiguity is so prevalent and so important as to alone justify the evolution of mental nodes.

Ambiguity can be said to occur when different nodes in the same domain simultaneously receive comparable levels of priming from bottom-up connections. Because studies to date have focused mainly on sentential ambiguities, I begin by discussing examples within the sentential system, but as we will see, ambiguity is a much more general and ubiquitous issue, applying to any node in any system.

Understanding the processing of ambiguity has been especially relevant for attempts over the past several decades to develop machines that can comprehend printed language. These efforts have often been thwarted by the fact that common words frequently allow ten or more distinct meanings (Kuno, 1967). Given this prevalence of ambiguity, what is surprising from the perspective of artificial intelligence is not that people *sometimes* experience difficulty with ambiguity (as shown in D. G. MacKay, 1966), but that they experience difficulty so rarely. The reason lies in the remarkable human capacity to resolve ambiguities by rapidly integrating different types of contextually specified information.

The Contextual Resolution of Ambiguity

Two basic characteristics of disambiguation must be explained in theories of perception: the extremely efficient use of context in disambiguation and the either–or resolution of ambiguity, which is the fact that we perceive either one interpretation of an ambiguous input, or the other, but not both at once (McClelland et al., 1986).

To illustrate the disambiguating effects of context, consider the ambiguous word *crane* and two of its meanings: *crane 1*, a bird with long legs, and *crane 2*, a mechanical hoist. Listeners quickly comprehend one meaning or the other for the word *crane* on the basis of context, including linguistic context, discourse context, situational context (specified via any sensory input or combination of inputs), and beliefs or general knowledge about the topic under discussion. The disambiguating contexts can either precede or follow the ambiguous word and can resolve the ambiguity within 700 ms (Swinney, 1979) or less (D. G. MacKay, 1970d). How can humans bring such large amounts of heterogeneous information to bear so quickly in resolving ambiguity? The contextual resolution of ambiguity constitutes a fundamental problem for artificial intelligence and for theories of speech perception alike.

Ambiguity and the Most-Primed-Wins Principle

A simple, multipurpose mechanism that can use any type of contextual information to automatically resolve any type of ambiguity is built into the node structure theory: the most-primed-wins principle. Figure 7.1 details the process for the ambiguous word *crane*, where a single phonological node, *crane*(stressed syllable), sends bottom-up connections to two lexical content nodes: *crane 1*(noun), which represents the meaning "a bird with long legs," and *crane 2*(noun), which represents the meaning "a mechanical hoist." This is a prototypical instance of ambiguity, because both *crane 1*(noun) and *crane 2*(noun) simultaneously receive comparable levels of priming from *crane*(stressed syllable) when listeners hear the word *crane* in isolation.

However, *crane 1*(noun) and *crane 2*(noun) are unlikely to achieve exactly equal priming in everyday speech perception, because a large number of additional (contextual) sources will be contributing additional priming to one node or the other. For example, *crane 1*(noun) will receive priming from contextual sources representing, say, visual perception of the bird, a discourse on cranes or on birds in general or a prior sentential context such as "The birds with the longest legs, biggest beaks, and greatest wingspan in swampy habitats are. . . ." In such contexts, *crane 1*(noun) will achieve greatest priming and become activated automatically under the most-primed-wins principle, rather than *crane 2*(noun), referring to the mechanical hoisting device (D. G. MacKay, 1970d).

Because disambiguating contextual information is almost invariably available during everyday conversations, the theory explains why ambiguity causes so little trouble in real life. Context normally ensures that the appropriate node receives more priming than any other node in its domain, which in turn ensures that the appropriate meaning becomes perceived. Ambiguity can only cause

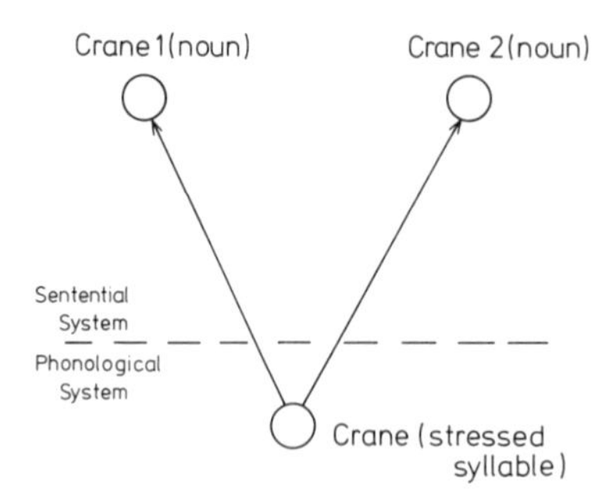

FIGURE 7.1. An illustration of ambiguity in the node structure theory. The syllable *crane* and two of its connected lexical content nodes are shown.

problems in the theory if the intended meaning is contextually inappropriate (as in garden path sentences, Chapters 2 and 4), or if both nodes representing the ambiguous meanings are receiving exactly equal priming, not just from below, but from all currently available contextual sources (D. G. MacKay, 1970d). Needless to say, these two conditions seldom arise in everyday life.

Conceptual Frequency and Context-Independent Disambiguation

Although normally sufficient, context is unnecessary for resolving ambiguities under the node structure theory, because the most-primed-wins principle can disambiguate words on the basis of conceptual frequency, even when the contextual cues that normally predispose perception of one meaning rather than the other are absent. Experiments in which an ambiguous word such as *crane* is presented in isolation nicely illustrate this point (Hogaboam & Perfetti, 1975). Here frequency of prior activation of the nodes representing the two meanings will determine which meaning becomes perceived, because frequency influences linkage strength, degree of priming, and probability of activation. Subjects will tend to perceive whatever meaning has higher frequency of occurrence in their personal experience. All other factors being equal, the ornothologist will perceive *crane 1*, whereas the hoist operator will perceive *crane 2*.

Averaged across subjects, of course, frequency of personal experience correlates with frequency in the language, and this explains a wide range of phenomena in the literature. To pick just one relevant example, subjects presented with a "ditropically ambiguous" expression, such as "He kicked the bucket," first perceive the high-frequency, idiomatic meaning, "to die" (Van Lanker & Carter, 1981). The literal (nonidiomatic) meaning, "a bucket was kicked," is less frequent than the idiomatic meaning and therefore less likely to become activated under the most-primed-wins principle.

The Either–Or Resolution of Ambiguity

The fact that ambiguities are resolved on an either–or basis imposes another basic constraint on theories of perception. Why do we initially perceive *only one* interpretation of the Jastrow rabbit-duck, or comprehend *only one* meaning of an ambiguous word or sentence (Kahneman, 1973; D. G. MacKay, 1966; 1970d)? In the example under consideration, why do we perceive either *crane 1* or *crane 2*, but never both simultaneously? Similarly, why do we comprehend an ambiguous sentence such as "They are flying planes" to mean roughly either, "Those machines are planes that fly!" or "Those people are in the process of flying planes!" but (almost) never *both* meanings simultaneously?

Unlike other theories, such as that of D. G. MacKay (1970d) and McClelland et al. (1986), the node structure theory requires no special-purpose mechanisms such as reciprocal inhibition between content nodes for accomplishing either–or resolution of ambiguity. The most-primed-wins mechanism, which is required for other reasons in activating each and every node, automatically resolves ambiguity in an either–or way. Under the most-primed-wins principle, only the most primed node becomes activated in a domain, including the domain of sentential sequence nodes. Thus, with presentation of an ambiguous sentence such as "They are flying planes," either COPULA or COMPLEX VERB becomes activated, and as a consequence, either *are*(copula) or *are flying*(complex verb) becomes activated, but not both at once (see McClelland & Kawamoto, 1986, for a similar account involving "case-role" units instead of sequence nodes or syntactic category units).

Of course, if instructed to do so, subjects *can* perceive first one and then the other meaning of an ambiguous word or phrase. However, perceiving the second meaning takes considerable time (D. G. MacKay & Bever, 1967), because a nonautomatic process is required to boost the priming of nodes representing the second meaning so that these primed but not activated nodes can become activated under the most-primed-wins principle when the activating mechanism is applied again. Interestingly too, the time required to perceive the second meaning is even longer when a *different* activating mechanism must be applied, as when the two meanings of the ambiguity belong to different domains or syntactic categories such as *like*(verb) versus *like*(preposition) (D. G. MacKay & Bever, 1967).

The Time Course of Disambiguation

Swinney's (1979) experiments provide a clear picture of the time course of disambiguation in sentence comprehension. Subjects listened to sentences such as "Rumor has it that for years, the government had been plagued with problems. The man was not surprised when he found several spiders, roaches, and other bugs in the corner of the room." Immediately after hearing the ambiguous word *bugs* in this passage, subjects saw either a word or a nonword on a screen they were watching, and made a yes–no lexical decision as quickly as possible. Swinney found that up to 400 ms following *bugs*, lexical decisions for words related to either of its meanings (e.g., *spy* and *ants*) were faster than for unrelated control words. The reason is that the node *bugs*(syllable) primes both *spybugs*(noun) and *insectbugs*(noun), thereby facilitating lexical decision time for *both* sets of semantically related words. However, *insectbugs*(noun) also receives contextual priming from *spiders*, *roaches*, and so on, and because *insectbugs*(noun) has more priming than *spybugs*(noun), it becomes activated under the most-primed-wins principle and enters conscious awareness about 700 ms after the syllable *bugs* is heard. Meanwhile, the original priming of *spybugs*(noun) has decayed, so that only lexical decisions for words related to the consciously perceived meaning (*ants*) receive facilitation 700 ms after *bugs*.

Ambiguity, Nonunique Priming, and Shades of Meaning

Ambiguity can be considered a special case of the more general phenomenon of nonunique priming, discussed in the previous chapter. Like ambiguity (Figure

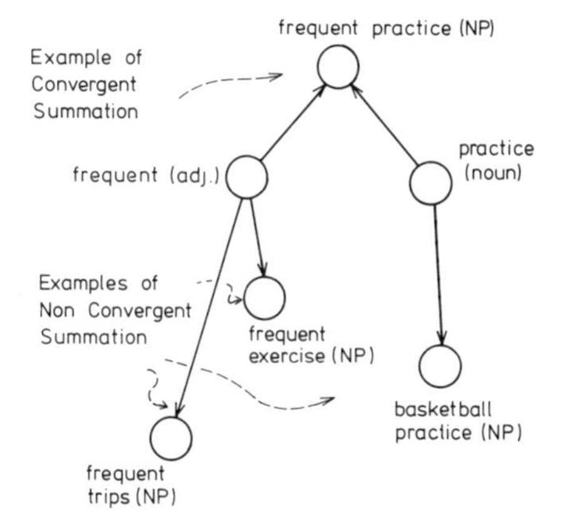

FIGURE 7.2. An illustration of nonunique priming, where four nodes in the same domain (Noun Phrase or NP) receive simultaneous first order priming. Note that priming converges or summates spatially for only one of these nodes receiving nonunique priming.

7.1), nonunique priming occurs whenever two or more nodes in the same domain become primed at the same time, usually from below. The difference is that in the case of ambiguity, nonunique priming arrives at comparable levels and from exactly the same number of bottom-up connections. To illustrate this difference, Figure 7.2 takes a typical (unambiguous) example of nonunique priming, the lexical input *frequent practice*. Activating the lexical nodes for *frequent* and *practice* primes a large number of noun phrase nodes at the same time. In particular, activation of *frequent*(noun) nonuniquely primes noun phrase nodes such as, say, *frequent trips*(noun phrase), and *frequent exercise*(noun phrase), as well as *frequent practice*(noun phrase) (Figure 7.2). Likewise, activation of *practice*(noun) nonuniquely primes noun phrase nodes such as, say, *basketball practice*(noun phrase), as well as *frequent practice*(noun phrase). However, because of convergent summation, *frequent practice*(noun phrase) will receive more priming than any of these other nodes in its (noun phrase) domain (Figure 7.2) and become activated automatically under the most-primed-wins principle, thereby determining perception. Clearly, ambiguity presents more of a problem for accurate comprehension than do other types of nonunique priming, but the most-primed-wins principle solves both problems in the same way under the theory, that is, automatically, without recourse to a conscious decision process and in a categorical or either–or way.

Interestingly, "shades of meaning" reflect the weaker case of nonunique priming under the node structure theory and therefore differ from ambiguity (unlike the proposal of McClelland & Kawamoto, 1986, p. 315, where ambiguity shades off seamlessly into shades of meaning). To illustrate shades of meaning, the *student* in *student of life* differs from the *student* in *medical student*. However, under the node structure theory, the same node represents *student* in these two examples. *Student*(noun) nonuniquely and nonconvergently primes both noun phrase nodes, *student of life*(noun phrase) and *medical student*(noun phrase). The differing shades of conceptual students are represented by the differing proposition nodes that *medical student* and *student of life* connect to.

Generality of the Problem

Recall that ambiguity can be said to occur in the node structure theory whenever different nodes in the same domain simultaneously receive comparable levels of priming from bottom-up connections. So defined, ambiguity can arise at every level in every system and may occur at least as frequently and can represent at least as much of a problem for phoneme recognition as for word comprehension (see also Massaro, 1981; McClelland & Elman, 1986). If segment nodes representing *g* and *k* receive comparable levels of bottom-up priming, the input can be said to be phonologically ambiguous between *g* versus *k*, for example.

Both experimental and theoretical considerations indicate that ambiguity is relatively common at the phonological and phonetic levels. Experiments such as D. G. MacKay (1978) show that, presented in isolation, words are highly ambiguous phonological entities and are subject to frequent misperceptions (Chapter 2). Moreover, if an easily perceived word in a naturally produced sentence is spliced out and presented out of context, subjects have difficulty telling what the word is (Cutler, 1985). Sentential context is apparently as necessary for deciphering speech sounds as for resolving lexical ambiguities, and the prevalence of phonetic and phonological ambiguities undoubtedly represents one of the main reasons why we have so far been unable to develop computer programs for providing accurate phonemic analysis of spoken English (see McClelland & Elman, 1986).

Theoretical considerations also suggest a prevalence of phonological ambiguities. Sensory analysis and phonological feature nodes generally prime a sizable set of segment nodes at the same time. Assume for the sake of illustration that a single node represents the phonological feature +voice (in either initial or final syllabic position). Activating this feature node would simultaneously prime the entire set of nodes representing voiced speech sounds, over 30 segment nodes in all. This, plus the occurrence of coarticulational overlap between adjacent segments further multiplies the theoretical likelihood of phonological ambiguity (McClelland & Elman, 1986).

Fortunately, however, direct resolution of phonological ambiguities is neither desirable nor necessary during everyday human sentence perception. It is not desirable because lower level disambiguation requires activation of phonological nodes, which would reduce rate of processing. If phonological nodes routinely became activated, the probability of activating the wrong node would also increase, causing perceptual errors (Chapter 4). And resolution is not necessary because the existence or nonexistence of nodes at higher levels usually resolves phonological ambiguities automatically. For example, an acoustic input halfway between *gastrointestinal* versus *kastrointestinal* is ambiguous at the phonological level between the syllables *gas* versus *kas*, between the segments *k* versus *g*, and between the features +voice versus −voice, but is unambiguous at the lexical level. Nonexistence of the node *kastrointestinal*(adjective) eliminates the ambiguity.

The principle of higher level activation reduces the probability of perceptual error in another way as well. Higher level nodes receive disambiguating information that is unavailable to lower level nodes but not vice versa. As noted previously in this chapter and in Chapter 4, lexical content nodes receive first-order priming from external (nonspeech) sources that cannot reach phonological nodes. A lexical content node, such as *apple*(noun), receives convergent priming from nonspeech sources representing, say, the smell or sight of an apple, as well as from phonological nodes representing the word *apple*. These external sources of priming can therefore serve to disambiguate an input at the lexical level but not at the phonological level. Phonological nodes representing the segments *l*, *p*, and *a* of the words *apples*, *pals*, or *laps* do not receive direct connections from the visual nodes representing *apples*, *pals*, or *laps*. And because phonological nodes are not subject to external sources of conceptual disambiguation, routine activation of phonological nodes (contrary to the principle of higher level activation) would further increase the probability of misperception.

Errors: A Cost of Integration

Mental nodes and the way they integrate information incur both costs and benefits. If integration of heterogeneous sources of information and automatic resolution of ambiguities represent benefits of mental nodes, errors represent a cost. In what follows, I review the general characteristics of errors discussed in previous chapters in order to show how automatic integration of heterogeneous sources of information via mental nodes contributes to errors in perception and action.

The Stroop effect illustrates how integration of different types of visual information can lead to errors (see also Norman, 1981), and it is an interesting (but often overlooked) historical fact that a whole range of Stroop-like effects were originally observed as a type of speech error. Meringer and Mayer (1895) reported that the names of colors and objects that a speaker is looking at, has heard spoken, or has recently read, often intrude as speech errors, substituting for a word that the speaker currently intends to produce. As is characteristic of speech errors in general, the intruding color words generally belonged to the same syntactic category (adjective), and subcategory (color adjective), as the word that the speaker intended to say at the time. Such errors illustrate how lexical nodes integrate priming that arises from a visual color, with priming that arises from reading, hearing, comprehending, and producing other (syntactically similar) words. Stroop errors occur because the wrong source of priming happens to dominate at the time when the activating mechanism is applied, so that the wrong lexical content node becomes activated under the most-primed-wins principle.

Blends and phonologically similar word substitutions also illustrate how errors arise from the integration of top-down and bottom-up priming during speech production. When speakers substitute words that are syntactically and phonologically similar, such as *a pressure* for *a present*, the syntactic similarity (both are nouns) can be characterized as a top-down effect, while the phonological similarity reflects a bottom-up effect (Chapters 2 and 6). "Freudian" errors also illustrate how mental nodes integrate top-down and bottom-up information of a much more heterogeneous sort during speech production (Dell, 1980; D. G. MacKay, 1982). An example is the substitution of *battle scared* for *battle scarred* in reference to an army officer whom the speaker believes is *scared of battle*. Although the speaker wishes to keep this opinion secret, top-down priming from this currently active "propositional belief" nevertheless automatically influenced which node in the (past participle) domain becomes activated. Another currently active belief also influenced the speaker's "correction" of the error, "battle scared, excuse me, I mean bottle scarred. . . ." Under the Freudian analysis (Freud, 1901/1914), this new error reflects an additional belief that this *battle scared* officer has been "*hitting the bottle*."

Freudian slips of the ear illustrate similar integrations of top-down priming (arising from propositional beliefs and attitudes) with bottom-up priming during ongoing word perception. An example is the misperception of *carcinoma* for *Barcelona* in the case of an individual who is temporarily concerned or preoccupied with this particular disease. The misperception occurs because priming for *carcinoma*(noun), arising from the preoccupation (top-down), and from aspects of the acoustic stimulus (bottom-up) exceeds priming for *Barcelona*(noun) arising from the input itself. As a consequence, the extraneous node *carcinoma*(noun) becomes activated under the most-primed-wins principle rather than the intended node *Barcelona*(noun).

In conclusion, errors in perception and action are largely attributable to the automatic manner in which mental nodes integrate priming from heterogeneous sources. However, I will argue in Chapter 9 that mental nodes also make errors especially easy to detect and correct, and this fact, together with the relative infrequency of errors in perception and action, suggests that errors constitute a small price to pay for the benefits of mental nodes, such as the automatic resolution of ambiguity.

8
Self-Inhibition and the Recovery Cycle

> Organization . . . is just as important a property of behavior as it is of perceptions. The configurations of behavior, however, tend to be predominantly temporal – it is the *sequence* of motions that flows onward so smoothly as the creature runs, swims, flies, talks, or whatever. What we must provide, therefore, is some way to map the cognitive representation into the appropriate *pattern* of activity.
>
> (G. A. Miller, E. Gallanter, K. H. Pribram, 1960, p. 11)

The present chapter describes a central concept in the node structure theory: the process of self-inhibition. I first examine the likely mechanism of self-inhibition and some theoretical reasons for postulating a self-inhibitory process. I then outline some evidence bearing on the self-inhibition hypothesis, some predictions that follow from self-inhibition, and an application of the self-inhibition hypothesis to the phenomenon of pathological stuttering.

The Process of Self-Inhibition

Self-inhibition is the inhibitory process that terminates the self-sustained activation of mental nodes and temporarily reduces their priming level to below normal or resting state. Like neurons, nodes have an absolute refractory phase, but it is of such brief duration (less than 1 ms) as to be irrelevant. For all practical purposes, self-inhibition introduces only a relative refractory phase in the excitability of mental nodes. Although self-inhibition reduces priming to below-normal levels, the node can still become activated if priming from other (external) sources is strong enough to meet the most-primed-wins criterion. Although self-inhibition is a built-in characteristic of all nodes (see the following discussion), only nodes that become activated rather than just primed during the course of perception and action exhibit self-inhibition. For example, because of the principle of higher level activation, low-level nodes do not exhibit self-inhibition during everyday perception and neither do "uncommitted" nodes (until they finally become activated and committed), even though the self-inhibition mechanism for already-committed nodes plays a role in the node commitment process (D. G. MacKay, 1987).

The Rationale and Nature of Self-Inhibition

Three types of factors call for a self-inhibition mechanism in the node structure theory: empirical factors, specific theoretical factors, and general ("Gestalty") theoretical factors having to do with overall simplicity and elegance of the theory (Fodor, 1980). However, I have not been able to fully develop the general theoretical rationale in the present chapter or even the present book. Self-inhibition forms part of simple and elegant mechanisms for processing self-produced feedback (Chapter 9) and for forming connections between nodes (D. G. MacKay, 1987). In this chapter I examine only the empirical and specific theoretical rationales for building the self-inhibition mechanism into the node structure theory.

The Specific Theoretical Rationale for Self-Inhibition

The theoretical necessity of something like self-inhibition has long been suspected (Feldman & Ballard, 1982; D. G. MacKay, 1969b). Neural activity tends to persist in the absence of inhibition, and unless components in a behavior sequence become inhibited after serving their function, general convulsion could result (Neumann, 1984). The node structure theory illustrates the theoretical necessity of self-inhibition in much more specific terms. Self-inhibition is needed to prevent disruptive effects of internal and external feedback on mental nodes. Internal feedback refers to the bottom-up priming that is transmitted to a superordinate node as soon as one of its subordinate nodes becomes activated. When a mental node becomes activated during production, it primes its subordinate nodes via top-down connections, thereby enabling activation. Activating these subordinate nodes then sends immediate internal feedback (priming) back to the superordinate node via the bottom-up connection required for perception. External feedback is less immediate and results from sensory analysis of the auditory or other perceptual consequences of an action, which likewise returns priming to the nodes that originated the action.

The internal and external feedback (priming) that results from the two-way (bottom-up and top-down) connections between most mental nodes (Chapter 2) leads to the possibility of reverberatory effects at virtually every level in a system, as illustrated by means of the hierarchically connected, but otherwise arbitrary, mental nodes in Figure 8.1, 1 (superordinate) and 2 and 3 (subordinate). During production, 1 becomes activated, and primes 2 via the top-down connection. However, when 2 and 3 subsequently become activated, this could lead to a reactivation of 1 because of the bottom-up connections required for perception involving these same nodes.

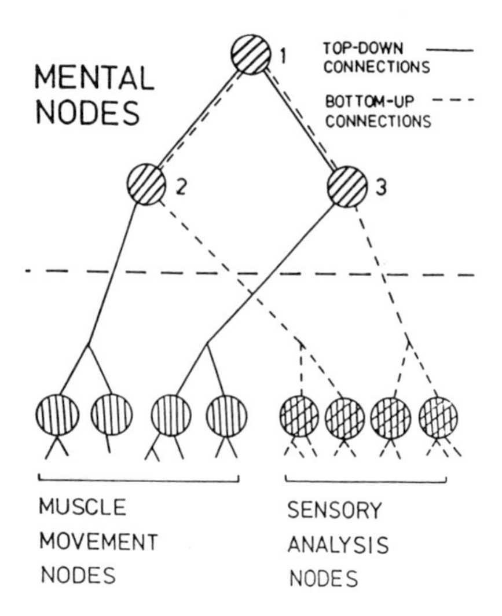

FIGURE 8.1. The relation between mental nodes, sensory analysis nodes, and muscle movement nodes.

Mental nodes must therefore become self-inhibited following activation to ensure that internal feedback (bottom-up priming) resulting from activation of subordinate nodes does not lead to repeated (reverberatory) reactivation of higher level nodes. If the self-inhibitory mechanism of a single mental node were to break down, an output resembling stuttering would result. If the self-inhibitory mechanism for large numbers of nodes were to break down simultaneously, the physiological and behavioral effects would resemble seizure or general convulsion.

Self-inhibition also helps to prevent similar and equally catastrophic effects of external feedback. For example, sensory analysis nodes automatically process the auditory feedback that arises during normal speech production and primes the same low-level mental nodes that generated the output in the first place, and reverberatory reactivation can only be prevented if the mental nodes receiving this returning, feedback-induced priming are still in their self-inhibitory phase following activation.

The Recovery Cycle

Self-inhibition is only the first stage in the cycle of recovery from activation. Two further stages are necessary to complete the cycle: hyperexcitability or postinhibitory rebound – a period during which self-priming rises above resting level (for the generality of postinhibitory rebound, see Grossberg, 1982) – and the final return to resting level.

In all then, four phases of excitability (summarized in Figure 8.2) follow multiplication of priming: activation, self-inhibition (the relative refractory phase), hyperexcitability, and the return to resting level of priming. The activation phase begins at time t_0 in Figure 8.2 and ends at time t_1, when self-inhibition

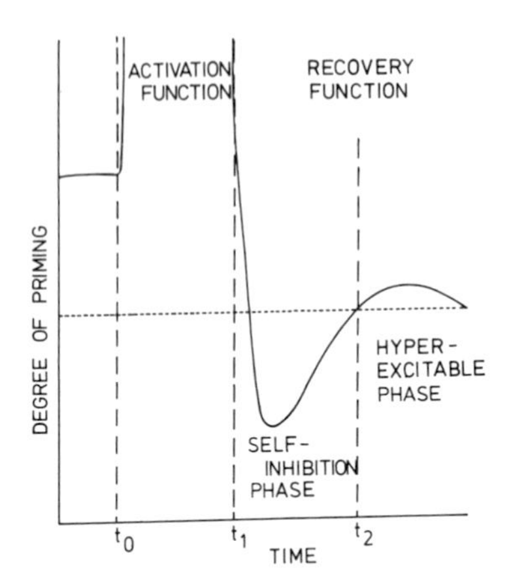

FIGURE 8.2. The activation and recovery phases for a single node. The activation function illustrates multiplication of priming beginning at time t_0, with self-sustained activation continuing until time t_1. The recovery function shows how priming first falls below resting level (self-inhibition) at termination of activation and then rebounds beginning at time t_2 (the hyperexcitability phase).

begins. This introduces the relative refractory phase, which lasts until t_2 in Figure 8.2 when priming first returns to resting level. The hyperexcitability or rebound phase follows, during which priming rebounds before returning to resting level again.

Temporal Characteristics of the Activation and Recovery Cycle

The duration of the entire activation and recovery cycle (from onset of activation until the final return to resting level) varies with the level of a node in the system. For the lowest level kinesthetic-muscle movement nodes, the entire cycle may last only a few milliseconds. However, for phonological nodes, the cycle can last up to 300 ms, with a relative refractory phase lasting as long as 100 ms, a hyperexcitability peak at 200 ms, and a return to resting level at 300 ms. For still higher level nodes, the activation and recovery cycle can take even longer.

Why must self-inhibition last so long in higher level (e.g., phonological) nodes, and why does the duration of self-inhibition vary with the level of a node in the system? One reason is that prolonged self-inhibition enables protection against external feedback, which takes a relatively long time to return in the case of mental nodes. A higher level node only begins to receive external feedback (priming) after three sets of events have taken place: (1) all of its subordinate nodes become activated, including its lowest level muscle movement nodes; (2) the environmental effects of the resulting action become relayed to the sensory receptors, enabling, for example, the airborne auditory feedback from speech to reach the ears; and (3) the sensory analysis and other subordinate nodes process the feedback and deliver priming to the higher level node(s) that originated the output. In order to prevent the reverberatory effects of external feedback, self-inhibition must therefore last as long as steps (1) to (3).

Why must the duration of self-inhibition vary with the level of the node in question? One obvious reason is that the time for steps (1) and (3) depend on the level of the node in an action hierarchy. The higher the node, the longer it will take for external feedback to return to that node. Another reason concerns internal feedback. A higher level node continues to receive internal feedback over a longer period of time than does a lower level node, so that again, self-inhibition must last longer for higher level nodes in order to prevent catastrophic reactivation.

Neural Mechanisms and the Recovery Cycle

In general, the basic properties of nodes (priming and activation) mimic by design the basic properties of neurons (potentiation and spiking) except for time scale and underlying mechanism, and the same is true of self-inhibition. The self-inhibitory process that follows activation of isolated neurons resembles the self-inhibition of mental nodes, but differs greatly in time scale. For example, the relative refractory period lasts 1 to 2 ms in isolated peripheral neurons, as compared to over 150 ms in self-inhibited mental nodes. Also unlike neural recovery, time characteristics of self-inhibition vary with the level of the node in question. The higher the node in a perceptual hierarchy, the longer the duration of self-inhibition. The mechanism underlying self-inhibition versus neural refractoriness also differs. Neural refractoriness reflects a physiochemical recovery process, whereas self-inhibition of mental nodes reflects complex neuronal interactions and evolved for reasons other than recovery per se.

What are the neural mechanisms underlying self-inhibition? The simplest possibility is that mental nodes consist of two interacting components: a parent neuron and an isolated inhibitory collateral or satellite neuron with an inhibitory connection to its parent. Like other isolated neurons, the inhibitory satellite has an absolute rather than a most-primed-wins threshold. Once its (extremely high) threshold potential has been reached, the satellite becomes activated (generates spikes), and once its potential falls below threshold, it deactivates. Thus, when the parent neuron has generated spikes over some prolonged period of time, the potential of its connected inhibitory satellite summates up to threshold, whereupon the satellite becomes activated and inhibits its parent neuron. This self-inhibition deactivates the parent neuron, thereby reducing the excitatory potential of its inhibitory satellite to below-threshold levels. As a result, the inhibitory satellite itself deactivates, enabling recovery from self-inhibition to begin in the parent neuron.

What determines the temporal parameters of self-inhibition in mental nodes? Although inhibitory satellites are a built-in characteristic of all mental nodes, the history of activation of a node determines when its self-inhibitory phase will begin. The process is described in Chapter 1 (and in Eccles, 1972). That is, the built-in connection between a parent neuron and its inhibitory satellite is extremely weak in the case of uncommitted (never previously activated) nodes, so that the parent neuron must remain activated for a very long time in order to build up sufficient potential to activate its inhibitory satellite, and thereby terminate activation of the parent neuron. However, as repeated activation increases the linkage strength of the connection between parent and satellite, the time required before onset of self-inhibition will automatically decrease. Because frequency of prior activation is directly related to the level of a node in the hierarchies for perception and action (Chapter 4; D. G. MacKay, 1982), the onset time of self-inhibition (i.e., the duration of activation) will therefore decrease with decreasing level of the node in question (all other factors being equal).

Empirical Evidence for Self-Inhibition and the Recovery Cycle

In what follows I discuss various sources of empirical evidence suggesting that mental nodes undergo a recovery cycle with the general characteristics previously discussed. The evidence ranges from neurophysiology, to errors in speech and typing, to repeated letter misspellings of dysgraphics, to the perception and recall of repeated letter misspellings by normal individuals, and even to the pattern of phoneme repetition in the structure of languages.

The Misspellings of Dysgraphics

A dysgraphic is someone who misspells common words with very high probability, due perhaps to cerebral injury, but not to lack of schooling or to a general inability to learn. The dysgraphic has a chronic spelling problem but sometimes spells one and the same word correctly on one occasion and incorrectly on others. And repeated letters often play a role in the misspellings. Lecours (1966) originally discovered the "repeated letter effect" in the misspellings of a dysgraphic (Lee Harvey Oswald) whose data were reanalyzed by D. G. MacKay (1969c) in a way that bears on the recovery cycle hypothesis. Oswald often dropped a repeated letter in a word, misspelling *elderly* as *eldery*, for example, but he sometimes added repeated letters of his own, as in *habitituated* and *Decemember*. D. G. MacKay (1969c) showed that deletions of repeated letters were significantly more common than additions and that deletions of the second of two repeated letters, as in *eldery* were significantly more common than deletions of the first, as in *ederly*, as would be expected if repeated letter deletions reflected self-inhibition of nodes that are to be activated in sequence.

However, D. G. MacKay's (1969c) most interesting data bearing on the recovery cycle hypothesis concerned the degree of separation of the repeated letters in Oswald's misspellings. Oswald frequently misspelled repeated letters that were close together, as in *anlyze*, but he rarely dropped repeated letters that were widely separated, as are the *i's* in *misspelling*. Figure 8.3 shows the probability that the dysgraphic *correctly* spelled words containing repeated letters as a function of the degree of separation of repeated letters in the sample at large. By way of illustration, the repeated *O's* in *cooperation* fall within the Separation 0 (zero) category because no letters separate the repeated *O's*. The repeated *A's* in *analyze* fall within the Separation 1 category because one letter separates the repeated *A's*, and so on, up to Separation 7.

FIGURE 8.3. The probability of correct spelling (versus letter deletion) as a function of the number of letters separating a repeated letter in the diary of Lee Harvey Oswald. This probability (PC) was calculated for each degree of separation as PC = Fd/F, where Fd is the frequency of repeated letter deletions such as *eldery* and F is the frequency of correct spelling. (From "The Repeated Letter Effect in the Misspellings of Dysgraphics and Normals" by D. G. MacKay, 1969c, *Perception and Psychophysics*, *5*, pp. 103–104. Copyright 1969 by Psychonomics Journals Inc., Austin, Texas. Reprinted by permission.)

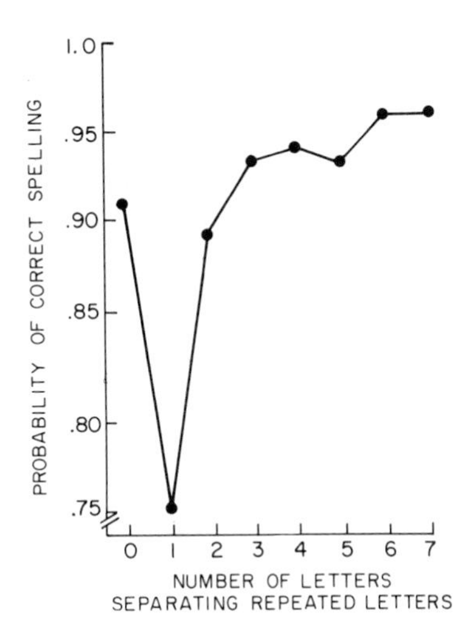

As can be seen in Figure 8.3, the probability of correctly spelling a repeated letter sequence was moderately high for immediately repeated letters (Separation 0), but dropped dramatically for repeated letters with Separation 1, and increased again to its highest level at Separation 6. The function in Figure 8.3 can be seen to resemble the recovery cycle function in Figure 8.2, except for the moderately high probability of correctly spelling immediately repeated letters. However, this special status of immediately repeated letters is to be expected. Unlike other repeated letters, immediately repeated letters generally are not pronounced in English, and do not represent separate phonemes. There are also special orthographic rules that apply only to immediately repeated letters and not to repeated letters with greater separation. Both of these factors suggest that data for repetitions with zero separation should be disregarded or treated separately in this and related functions.

The Perception and Recall of Misspellings

Another source of evidence for the recovery cycle hypothesis comes from a study of the detection of experimentally constructed misspellings. D. G. MacKay (1969c) examined how readily normal subjects could perceive and recall misspellings that resembled those produced by the dysgraphic discussed previously. The misspellings were planted in sentences that subjects read at a rate of about 77 ms per letter. The subjects then attempted to recall the sentence, writing it out exactly as it was spelled, guessing at misspellings if necessary. Following recall, each sentence was presented again, and subjects responded yes or no to each spelling error in turn, depending on whether they had noticed the error when reading the sentence.

The results showed that normal individuals experienced greatest difficulty in perceiving and recalling those experimentally constructed misspellings that resembled the ones produced most frequently by the dysgraphic. Repeated-letter misspellings such as *eldery* were harder to perceive than nonrepeated-letter misspellings such as *eldely*. Similarly, given that the subject claimed to perceive an error when reading the sentence, misspellings that involved repeated letters were more difficult to recall than those that did not.

As expected under the recovery cycle hypothesis, the functions relating degree of separation of the repeated letters to both perception and recall resembled the dysgraphic function in Figure 8.3. The perception function appears in Figure 8.4. The solid line in Figure 8.4 plots the probability of correctly perceiving nonrepeated-letter misspellings, while the broken line plots the probability of correctly perceiving repeated-letter misspellings, as a function of the degree of separation of the repeated letters. As can be seen in Figure 8.4, repeated-letter misspellings were only harder to detect than nonrepeated-letter misspellings when zero to three letters separated the repeated letters. With more than three intervening letters, repeated-letter errors were as easy or easier to detect than nonrepeated-letter errors.

The recall function appears in Figure 8.5, which plots the probability of correctly recalling a misspelling that was correctly perceived. The solid line in Figure 8.5 represents recall for nonrepeated-letter misspellings, while the broken line represents recall for repeated-letter misspellings, as a function of the degree of separation of the repeated letters. As can be seen in Figure 8.5, repeated-letter misspellings were only harder to recall than nonrepeated-letter misspellings when zero to two letters separated the repeated letters. With more than three intervening letters, repeated-letter errors were as easy or easier to recall than nonrepeated-letter errors.

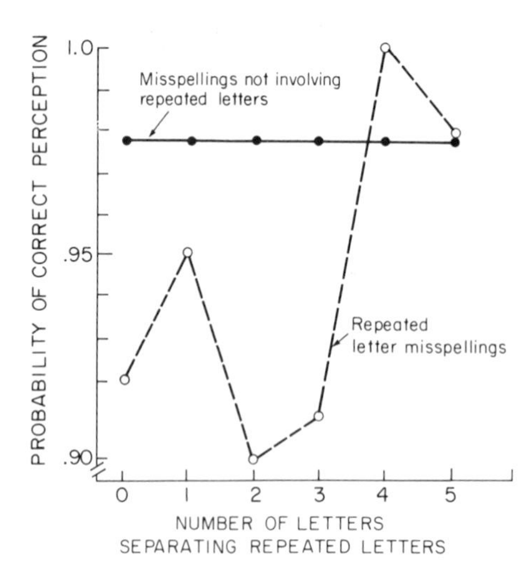

FIGURE 8.4. The probability of perceiving repeated-letter misspellings such as *eldery* (broken line) and nonrepeated-letter misspellings (solid line) such as *eldely*. (From "The Repeated Letter Effect in the Misspellings of Dysgraphics and Normals" by D. G. MacKay, 1969, *Perception and Psychophysics*, 5, pp. 103–104. Copyright 1969 by Psychonomics Journals Inc., Austin, Texas. Reprinted by permission.)

FIGURE 8.5. The probability of recalling repeated-letter misspellings (broken line) and nonrepeated-letter misspellings (solid line). (From "The Repeated Letter Effect in the Misspellings of Dysgraphics and Normals" by D. G. MacKay, 1969, *Perception and Psychophysics*, *5*, pp. 103–104. Copyright 1969 by Psychonomics Journals Inc., Austin, Texas. Reprinted by permission.)

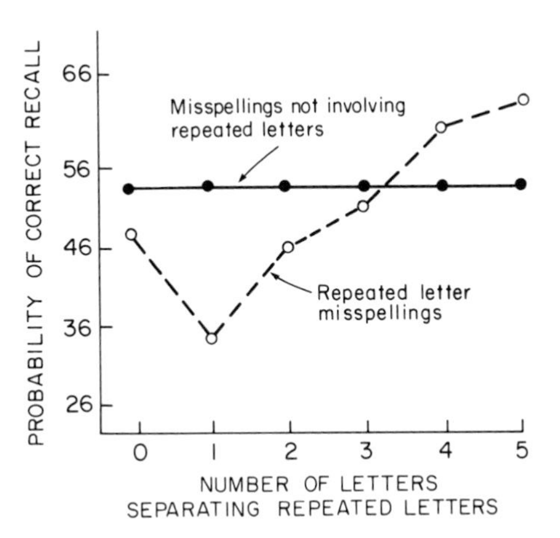

An additional prediction from the recovery cycle hypothesis concerned the nature of repeated letter misspellings. If a repeated letter has been *added*, as in *elderdly*, then the added letter should be easier to perceive with high degrees of separation (4 to 5), than with low degrees of separation (0 to 2). However, if a repeated letter has been *deleted* as in *eldery*, then the error should be as easy to perceive with low degrees of separation (0 to 2) as with high (4 to 5), because of the hypernormal excitability of the node for the expected letter at the later point in time.

LIMITATIONS OF D. G. MACKAY (1969c)

Although D. G. MacKay's (1969c) results support the recovery cycle hypothesis, there are a number of ways to extend and refine these observations using computer-controlled stimulus presentation. Consider, for example, a task where subjects must detect a set of possible target letters in a rapidly presented sequence of letter displays, as in Shiffrin and Schneider (1977), except that target letters can be repeated in the sequence. The subject has three tasks. The top priority task is to respond as quickly as possible to the occurrence of any member of the set of targets, and the other, subsidiary tasks are to indicate (1) whether a second target followed the first and (2) the identity of the target(s). In the special case where the *same* target letter is repeated (in either identical or different locations in the display), the node structure theory predicts that detecting the first target presentation will (1) interfere with detection of the second when the intervening interval is short (up to 50 ms, say) but will (2) facilitate detection of the second when the intervening interval approximates some critical value corresponding to the hyperexcitability peak. Specifying this critical value in advance is difficult, because the recovery cycle varies with the level of a unit and its prior history of activation, but a value of 200 ms can be expected if letter-detection nodes resemble those for phonemes.

Repeated Phonemes in Fast Speech

When speaking rapidly, speakers tend to drop immediately repeated phonemes, as might be expected under the recovery cycle hypothesis. However, determining exactly which nodes are responsible for this phenomenon is difficult. For example, when speaking rapidly, speakers often drop one of the immediately repeated *k's* in *take care* (Heffner, 1964), but it is difficult to specify which *k* was dropped, or even to determine whether the omission was intentional. Moreover, such omissions cannot be attributed to self-inhibition at the segment level, because two different segment nodes control production of these *k's*, *k*(final consonant group) and *k*(initial consonant group), and because self-inhibition of nodes below the segment level could cause these omissions.

Errors in Speech and Typing

The fact that anticipations are much more common than perseverations in everyday speech errors (Cohen, 1967) bears directly on the self-inhibition hypothesis. So does Type I motor masking. Unlike the omissions that occur in rapid speech, the class of speech errors that D. G. MacKay (1969b) labeled Type I motor masking is undeniably inadvertent. The speaker unintentionally drops the second of two segments due to be repeated in close succession. An example from Meringer and Mayer's (1895) corpus of German errors is *der iese* instead of *der Riese*, where the immediately repeated /r/ has been dropped. Because Type I motor masking appears to be rare and requires that an observer make subtle perceptual discriminations under less than optimal observational conditions (Cutler, 1982), these errors only weakly support the self-inhibition hypothesis (D. G. MacKay, 1969b), and warrant further investigation using experimental error-induction procedures.

However, omission errors in skilled transcription typing exhibit a similar phenomenon that cannot be attributed to observer error. High-speed videotapes of the finger movements of skilled typists indicate that typestrokes exhibit three phases (Grudin, 1983): (1) a movement that positions the finger over the key, (2) a rapid downstroke to strike the key, (3) and an equally rapid rebound or lift-off from the key. During an omission error, the finger either fails to move toward the key, or fails to execute the downstroke. Like Type I masking errors in speech, typestroke omissions are usually preceded by an identical letter or identical finger movement, either in the same word or in the immediately preceding word. For example, Grudin (1983) had skilled typists transcribe a text as rapidly as possible and found that half of the typists had omitted the third *i* in the word *artificial*. Moreover, in Grudin's overall sample of omission errors, this left-to-right masking (omission) effect was strong enough to override a general tendency for word-initial letters to be correctly typed. Examples of such word-initial omissions are *the ntire* for *the entire*, and *keep utting* for *keep putting*. These typestroke repetition errors suggest that following activation, the mental nodes controlling skilled typestrokes undergo a period of self-inhibition that interferes with the subsequent execution of an identical typestroke due to follow closely in time (see also Grudin, 1983). Omissions of immediately repeated movements required for different letters suggest a similar process for nodes that perceive and/or produce movement components, rather than whole letters.

On the surface, omission errors in typing, and the recovery cycle hypothesis in general, seem to conflict with the fact that people can repeat a finger movement more rapidly with a single finger than with two adjacent fingers in alternation. Why is the maximal rate of key pressing faster with one finger than with two fingers? One reason is that repeated finger movements can be executed entirely within the muscle movement system. Mental nodes are unnecessary for producing an up-and-down movement of the finger, so that the temporal parameters of self-inhibition in mental nodes are irrelevant to these movements. Another reason is that two-finger movements require a sequencing process that is not required for single-finger movements. The additional time required for sequencing two-finger movements (Chapter 3) could greatly reduce the maximum response rate.

Electrophysiological Evidence

Although the figure of 200 ms for the hyperexcitability peak of mental nodes is only approximate, it receives support from the electrophysiological literature. For example, postinhibitory rebounds have been observed that follow activation by as much as 200 ms in central neuronal aggregates (see Chang, 1959; Grossberg, 1982; Martin, 1985). These central neuronal aggregates therefore differ from isolated peripheral neurons, where the hyperexcitability peak arrives orders of magnitude sooner.

Electromyographic Evidence

Electromyographic potentials without full-blown muscle movements occur during mental practice and indicate in the theory (D. G. MacKay, 1981) that the relevant muscle movement nodes are being primed, and in some cases inadvertently activated. Because priming is assumed to peak spontaneously at about 200 ms following self-inhibition, the theory therefore predicts that the appropriate muscles will exhibit a peak in electromyographic potentials about 200 ms after production of a segment during speech production. Evidence for such a peak is found in a study by Ohala and Hirano (1967). They had subjects produce the syllable *pa* while recording electromyographic potentials in the obicularis oris muscles of the lips, and they observed two peaks of electromyographic activity. The first was causally related to contraction of the lip muscles, and the other, smaller peak about 200 ms later, suggests support for the recovery cycle hypothesis.

Phoneme Repetition in the Structure of Languages

Somewhat indirect but nevertheless interesting support for the recovery cycle hypothesis comes from a study of phoneme repetition in the structure of words (D. G. MacKay, 1970c). The rationale for the study was as follows. If a recovery cycle determines how easy it is to repeat an element in a word, then the structure of repeated elements in words should reflect that recovery cycle. Immediate repetition of an element should be rare, reflecting the self-inhibitory phase of the recovery cycle, but at some point following self-inhibition, repetition should become more likely than would be expected by chance, reflecting the hyper-excitability phase of the recovery cycle. Applying this hypothesis to phonemes, if a recovery cycle constraint makes immediate repetition of phonemes difficult, then the phonological structure of words in many languages should reflect this constraint, so that immediate repetitions of phonemes should be rare. Similarly, if phoneme repetition after some interval is extremely easy, then repetition of phonemes in the structure of words should be highly likely after a corresponding degree of separation. Implicit in this rationale is the (uncontroversial) assumption that speakers don't normally alter their rate of speech during ongoing production of words containing repeated phonemes.

To test this recovery-cycle prediction, D. G. MacKay (1970c) examined patterns of phoneme repetition in words of various lengths in two very different languages, Serbo-Croatian and Hawaiian. A very large and random sample of words was examined, and the probability of segment repetition at the various separations was corrected on the basis of word length, because long separations for short words are impossible.

Figure 8.6 shows the pattern of phoneme repetition for both languages, averaged over consonants and vowels. As can be seen, immediate repetition of phonemes was highly unlikely for both languages and significantly less likely than would be expected if phoneme repetition in a word were a random event (represented by the dashed lines in Figure 8.6). This finding is interesting because immediate repetition of phonemes is clearly physiologically possible. Sequences such as /b-b-b/ are a common occurrence in stuttering (Chapter 10).

With wider degrees of separation, probability of repetition rose sharply, peaking with three intervening elements in both Serbo-Croatian and Hawaiian (Figure 8.6). These peak repetition probabilities exceeded what would be expected if phoneme repetition in a word in either language were a random event. Following the peak, probability of repetition declined to approximately chance level. However, there was a difference between vowels and consonants (D. G. MacKay, 1970c), reflected in part by the two initial peaks in the functions of Figure 8.6. In both Serbo-Croatian and Hawaiian, probability of *vowel* repetition peaked with one intervening element, whereas probability of *consonant* repetition peaked with three intervening segments.

Differences between the two languages were superimposed on the similarities. As can be seen in Figure 8.6, the repetition pattern was more erratic for Hawaiian than for Serbo-Croatian. The reason is that virtually all of the Hawaiian syllables

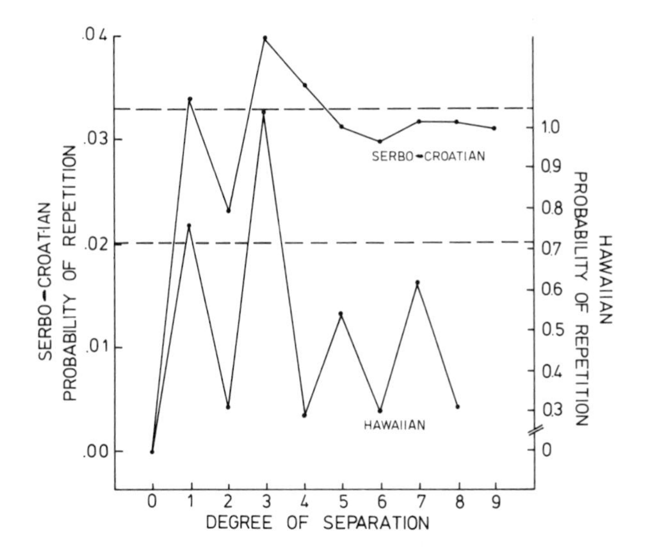

FIGURE 8.6. The probability of phoneme repetition in Serbo-Croation (left ordinate) and Hawaiian (right ordinate) as a function of degree of separation of the repeated phonemes. The horizontal lines represent probabilities under the hypothesis that phoneme repetition is a random event.

had a simple consonant-vowel structure, whereas Serbo-Croatian syllables were much more variable and complex in structure. Thus, because vowels and consonants alternate in Hawaiian, segment repetition at even numbered degrees of separation was of necessity infrequent. Given some way of factoring out syllable structure in computing the probabilities of repetition, the Hawaiian and Serbo-Croation functions would have resembled each other more closely.

As D. G. MacKay (1970c) points out, the peaks in Figure 8.6 suggest a "law of latent alliteration," and differences between Hawaiian versus Serbo-Croatian syllable structures serve to illustrate the statistical nature of that law. Many other factors play a role in determining the phonological structure of words and can thereby override the law of latent alliteration in any one instance. Another important factor, especially for complex or derived words, is the nature of affixes and stems that combined to make up the word. For example, adding a different prefix could change the structure of repeated phonemes in the word. It is therefore remarkable that a consistent repetition structure should appear at all and that it should be so similar for Hawaiian and Serbo-Croatian, languages that differ in type of segments, number of segments (13 versus 35), and average number of segments per syllable and per word. The data therefore suggest, but do not prove, that the pattern of segment repetition in Figure 8.6 is representative of all languages, as would be expected if general properties of nervous action were an underlying cause. In theory of course, a similar pattern should appear with only minor variations in all languages.

Phoneme Repetition and Language Evolution

Phoneme repetition is a factor in the phonological changes that words undergo over time, because repeated segments often become dropped as a language evolves. An example is the Latin word *stipipendium*, which changed to *stipendium*, dropping the repeated /p/ and repeated /i/ (Meringer & Mayer, 1895). Such changes fit the prediction that phoneme repetition represents a factor in the evolutionary processes that words undergo in the history of a language, but the extent to which the hypothesized recovery cycle plays a role in the initial invention of words, or in the evolutionary changes that occur over time, is currently unknown. A great deal of further research is required to fully specify how the pattern of segment repetition within words evolves.

Limits of the Recovery Cycle Hypothesis

I come now to some limits of the recovery cycle hypothesis for theories of speech and other behaviors. One limit is that the constraints on segment repetition discussed previously only apply to nodes that receive parallel top-down and bottom-up connections and therefore require protection from internal feedback. For example, the recovery cycle hypothesis does not apply to the lowest level mental nodes in an action hierarchy, which in the case of speech, are distinctive feature nodes. In theory, distinctive feature nodes are undisturbed by internal feedback, because they do not receive parallel top-down and bottom-up connections. It is therefore interesting that immediate repetition of distinctive features is common in many languages, and the period during which distinctive feature nodes remain activated generally exceeds that of segment nodes, spanning several adjacent segments, which could not occur if distinctive feature nodes undergo self-inhibition.

Another limit to the recovery cycle hypothesis concerns the potential or maximal rate of segment repetition. At best, data on the structure of segment repetition in a language only suggest a repetition pattern that may be *easily* produced at average rates for everyday speech. The data say nothing about the potential or maximal rate for producing syllables or for moving the speech muscles. Self-inhibition introduces a relative rather than an absolute refractory phase. By exerting greater effort, speakers can greatly exceed the natural repetition rates suggested in Figure 8.6.

Yet another limit to the recovery cycle hypothesis concerns the type of skill being investigated. Similar constraints on repetition rates can only be expected for skills that employ shared input–output nodes. Skills that do not engage mental nodes for production and perception will not display the same recovery cycle pattern.

Finally, the self-inhibition hypothesis applies only to highly practiced and rapidly activated mental nodes and not to seemingly similar phenomena that occur in unskilled behavior, such as psychological refractoriness and the Ranschburg (1902) effect, discussed in the following section.

Self-Inhibition and the Psychological Refractory Effect

The psychological refractory effect refers to the fact that responding to one signal can prolong the time required for responding to a subsequent signal. The basic empirical paradigm is as follows. A signal such as a tone is presented, to which the subject must rapidly respond with, say, a key press. Then, before the key press has been made, a second stimulus such as a light appears, to which the subject must make a different response, say, a vocal "yes." The remarkable and frequently replicated result is that responding to the first signal lengthens reaction time to the second. In the present example, the tone lengthens reaction time to the light, relative to when the light alone is presented. The difference in reaction time between these two conditions is known as the refractory delay and typically exceeds 200 ms. Refractory delays diminish over time, but often exceed 100 ms even when the response to the first signal has been completed before onset of the second signal.

On the surface, the psychological refractory effect seems to resemble effects of self-inhibition discussed above. Both effects have a central origin, are very general in nature, and reflect an inhibitory process. As Keele (1973) points out, psychological refractory effects occur with many input–output skills and reflect central rather than purely sensory or purely motor limitations. Refractory delays remain when different perceptual systems process the signals and when different output systems execute the responses, as in the example previously cited.

However, only some of the refractory delays in the literature can be attributed to self-inhibition of content nodes. When both stimuli and both responses differ in the psychological refractory paradigm (again as in the above example), different content nodes are becoming activated and self-inhibited, so that self-inhibition cannot explain the interference. However, the interfering stimuli and responses in the psychological refractory paradigm are almost invariably similar, so that their content nodes probably belong to the same domain. This suggests that psychological refractory effects may hinge on the reactivation of an activating mechanism or sequence node which has recently been quenched (Chapter 1). Under this hypothesis, refractory delays should disappear following extensive practice with independent input–output mappings. The practice will enable both mappings to engage nodes in different domains, activated via different sequence nodes (D. G. MacKay, 1987).

This view explains two otherwise puzzling sets of conflicting results in the psychological refractory literature. One concerns the effects of practice on psychological refractoriness. Some studies have shown that refractoriness disappears with practice (e.g., Greenwald & Schulman, 1973), while others have shown that it does not, even after 87 practice sessions (e.g., Gottsdanker & Stelmach, 1971). Under the theory, these conflicting outcomes reflect differences in compatibility of the input–output mappings that the tasks engage. Greenwald and Schulman's (1973) tasks involved highly compatible or practiced input–output mappings, specifically, moving a lever in the direction of an arrow and shadowing (naming auditorily presented letters). These highly practiced mappings eliminated the psychological refractory effect. Responding to one stimulus had no effect on the rate of responding to the other stimulus. Under the node structure theory the reason for this absence of effect is that these different and highly practiced stimulus–response mappings engaged independent activating mechanisms, eliminating the possibility of cross-talk effects.

In Gottsdanker and Stelmach's (1971) tasks on the other hand, the input–output mappings were nonindependent, so that practice could not eliminate the possibility of cross-talk effects. That is, both stimuli were visual, and both responses were manual and engaged nodes in the same domains (perhaps even some of the same nodes in the same domains). As a result, identical activating mechanisms and content nodes were required for both responses, causing delays attributable to quenching, self-inhibition, and speed–accuracy trade-off.

The node structure account of psychological refractory effects also explains another set of conflicting results. When subjects are instructed to simply observe rather than respond to the first stimulus, some studies (e.g., Rubinstein, 1964) have reported full-blown refractory delays, whereas other studies (e.g., Borger, 1963) have reported no delays whatsoever. The difference is attributable to the fact that observation alone only causes refractory delays under the node structure theory when the principle of higher level activation must be abandoned and lower level nodes must become activated, that is, when the two signals are unfamiliar or highly similar and difficult to distinguish (D. G. MacKay, 1987). For example, Rubinstein (1964) observed full-blown refractory effects—even though the first signal required no response whatsoever—when the two stimuli were highly similar visual forms presented to the two eyes or were highly similar noises presented to the two ears. However, Rubinstein (1964) also showed that these refractory delays disappeared when the noise was presented to one ear as the first stimulus and the visual form was presented to the opposite eye as the second stimulus, or vice versa. In this condition, the stimuli occupy different domains, and the first stimulus neither activates nor primes nodes in the same domains as are required for producing the second response. As a result, the first stimulus interferes with neither perception of the second stimulus, nor production of the second response.

Finally, it should be noted that some refractory effects call for an explanation in terms of encoding processes, rather than either quenching or self-inhibition. For example, it is often the *first* response that is slowed down when the second signal arrives soon after the first. Moreover, with very short interstimulus intervals and prior experience with the stimuli, their order, and their interstimulus interval, the two responses sometimes become produced in quick succession and without refractory delays (Welford, 1968). As others have noted, the subjects can encode the stimuli as a pair and produce the responses as a group under these conditions, thereby avoiding refractory effects.

Self-Inhibition and the Ranschburg Phenomenon

The Ranschburg phenomenon refers to an inhibitory effect of repeated items on the immediate recall of short sequences of items. Sequences containing repeated items are recalled more poorly than sequences containing nonrepeated items. In his original (1902) demonstration, Ranschburg presented a simultaneous string of six digits for about 333 ms, and found that a string containing repeated digits was more poorly perceived or recalled than a string containing nonrepeated digits. Moreover, repeated digits, and especially the second of two repeated digits, were the ones most poorly perceived or recalled. Ranschburg attributed these findings to a general inhibitory process for repeated items and predicted analogous effects in spelling and writing.

The Ranschburg effect has stimulated a large number of experiments over the last 85 years, and these more recent experiments suggest that poor performance on repeated items is attributable to a complex conspiracy of factors, rather than just inhibition per se (Jahnke, 1969). For example, encoding factors play a big role in the Ranschburg paradigm. Subjects sometimes recall that an item was repeated but can't recall which one and inadvertently repeat the wrong one. Even when they accurately recall which item was repeated, subjects often place the repeated items in the wrong positions in the overall sequence. Finally, *immediately* repeated elements generally are perceived as a group and are recalled *better* than nonrepeated items. All of these findings are attributable to encoding effects (the process of forming connections between nodes in the node structure theory) and are irrelevant to the activation and self-inhibition of already formed nodes and connections.

Moreover, some of the encoding processes in the Ranschburg (1902) paradigm are rather artificial and fundamentally unlike encoding processes for, say, natural speech production (Jahnke, 1969). For example, effects of spacing on recall of repeated elements in the Ranschburg paradigm sometimes resemble recovery cycle effects in speech, but only superficially. *Facilitatory* rather than inhibitory effects are observed in widely separated elements but these effects vary with position in the list, spacing per se, and ease of encoding (e.g., repeated items at the beginning and end of a list are especially easy to encode and recall; Jahnke, 1969). The Ranschburg paradigm also allows guessing strategies, which can account for the poorer recall of repeated items, because guesses have a lower probability of being correct when items from a limited set (such as the digits 0 to 9) are repeated than when they are not repeated (Hinrichs, Mewalt, & Redding, 1973). Together these factors make the Ranschburg paradigm a poor choice for testing the self-inhibition hypothesis.

Self-Inhibition and Pathological Stuttering

Having reviewed the evidence relating to self-inhibition and the recovery cycle, I now apply the recovery cycle hypothesis to the phenomenon of pathological stuttering. I begin with an overview of similarities and differences between different types of pathological stuttering, and then compare stuttering with other types of speech errors.

Types of Pathological Stuttering

Stuttering is a complex disorder that falls into two general categories: feedback-induced stuttering and intrinsic stuttering. The latter is not attributable to a disruptive effect of feedback but to the motor control process per se. I discuss the evidence for distinguishing between these two categories of stuttering in Chapter 10. Interestingly, feedback-induced and intrinsic stuttering have been responsible for a major division within the field itself. Two major approaches to the study of stuttering have gone their separate ways over the past several decades, one viewing stuttering as a disorder in motor control, and the other as a disorder in the processing of auditory feedback (Garber & Siegel, 1982).

I begin with a brief description of the general characteristics shared by both feedback-induced and intrinsic stuttering and then concentrate on intrinsic stuttering in the remainder of the chapter, leaving feedback-induced stuttering for Chapter 10.

General Characteristics of Pathological Stuttering

The motor control problem in pathological stuttering can be separated into an unlearned component and a learned component. All stutterers exhibit the unlearned component, the speech errors known as repetitions, prolongations, and blocks. In repetition errors usually only single consonants or consonant clusters are repeated, and only occasionally a syllable or monosyllabic word (Van Riper, 1982). Prolongations involve the unbroken lengthening of a (continuant) phoneme. For example, a prolongation within the word *practice* might lengthen the /*r*/ to more than 40 times its normal duration. Descriptively, it is as if the articulators become "locked" in position, resulting in prolonged production of a continuant sound. Blocks reflect an inability to utter any sound at all and occur most often at the beginning of words and utterances. As Van Riper (1982) pointed out, blocks can be considered a special type of prolongation where one or more articulators (e.g., the velum, lips, or glottis) become locked in an obstructive position, preventing speech by virtually eliminating airflow.

The other, learned component of stuttering reflects the stutterer's ability to recognize and anticipate the occurrence of repetitions, prolongations, or blocks and attempt to avoid them or shorten their duration by, for example, contortion of facial muscles, changes in breathing pattern, or altered choice of words. These behaviors are characterized as learned because they don't appear in children who stutter or in normal adults who occasionally repeat, prolong, or block on speech sounds (Meringer & Mayer, 1895). Children and normal adults haven't developed anticipatory fears of these errors and therefore don't exhibit the learned attempts to avoid them.

Stuttering and Other Speech Errors

The present book focuses on the unlearned component of stuttering: repetitions, prolongations, and blocks. Stutterers sometimes make other errors, and nonstutterers sometimes stutter (Meringer & Mayer, 1895). The difference is that stutterers make repetitions, prolongations, and blocks orders of magnitude more frequently than do nonstutterers but make other errors with normal frequency (D. G. MacKay & Soderberg, 1972). In what follows, I compare stuttering with other speech errors in greater detail, my ultimate goal being to explain stuttering within the same framework as other speech errors.

Similarities Between Stuttering and Other Speech Errors

Stuttering resembles other speech errors in several respects. Variables such as ambiguity influence stuttering and other speech errors in similar ways (See D. G. MacKay, 1969a). Also, as in other errors, stuttering decreases with practice in producing a sentence (the so-called adaptation effect, Brenner, Perkins, & Soderberg, 1972) and with reduction in rate of speech (Perkins, Bell, Johnson, & Stocks, 1979). These parallel effects of practice and speech rate on stuttering and other speech errors are readily explained within the node structure theory (D. G. MacKay, & MacDonald, 1984).

Stuttering and other speech errors also occupy similar phonological loci. Like other errors such as anticipations, stuttering often occurs at the beginning of words and sentences (a phenomenon attributable to anticipatory priming) and involves stressed syllables more often than unstressed syllables (D. G. MacKay, 1970a). Finally, stuttering is not limited to people classified as stutterers. Meringer and Mayer (1895) showed that "nonstutterers" also produce repetitions, blocks, and prolongations, although subsequent collections of speech errors (e.g., Fromkin, 1973) have by and large excluded these everyday errors. The Epilogue following Chapter 10 examines the scope, rationale, and effects of this exclusionary strategy.

Differences Between Stuttering and Other Speech Errors

One of the main differences between stuttering and other speech errors, its individual-specific and error-specific nature, has already been noted. The same individuals produce repetitions, prolongations, and blocks with much higher frequency than normal speakers, but make other types of errors such as spoonerisms with about the same frequency as normal speakers. Another difference is that adult stutterers can anticipate when they are going to stutter, whereas nobody anticipates or struggles to avoid making other errors.

Stuttering also differs from other errors in disruptiveness or severity. Sometimes stuttering disrupts communication for sustained periods of time and cannot be voluntarily "corrected," unlike other errors, which either pass unnoticed, or are quickly and easily repaired. Also unlike other speech errors, stuttering is not always limited to the speech-motor system but manifests itself in other activities of stutterers. For example, stutterers have more difficulty than nonstutterers in reproducing the timing of a sequence of syllables or finger taps (M. H. Cooper & Allen, 1977), as if both speech and finger movement shared the same timing deficit. Stuttering also relates to the processing of sensory feedback in a way that other errors do not (see Chapter 10). Finally, as discussed below, stuttering and other errors seem to originate at different levels of speech production.

The Level at Which Stuttering Originates

Although the level at which stuttering originates has been a source of controversy for the past half century (see D. G. MacKay & MacDonald, 1984), current evidence suggests that stuttering can be localized within the muscle movement system. I review some of these lines of evidence in the following discussion.

Fluency During Internal Speech

Stutterers do not report stuttering during internal speech, a phenomenon that is sometimes attributed to the possibility that speaking to one's self may provoke less anxiety and therefore less stuttering than speaking to others. However, reduced anxiety cannot fully account for the absence of stuttering during internal speech, because stutterers frequently report stuttering when speaking *aloud* to themselves (Van Riper, 1982).

By way of contrast, other speech errors occur as often during internal speech as during overt speech. Dell (1980) provided the crucial evidence. His (normal) subjects repeated tongue twisters (e.g., *Unique New York*) either aloud or to themselves at a fixed but rapid rate and reported the occurrence of errors. The results showed that other errors (e.g., transpositions, perseverations, and anticipations as in misproduction of *New York* as *You Nork*, *New Nork*, or *You York*) occurred with exactly the same frequency during both internal and overt speech. Under the node structure theory, this finding suggests that these other errors can be localized within the phonological system, which is shared by internal and overt speech, rather than within the muscle movement system, which is not. And, because stuttering occurs during overt, but not during internal speech, stuttering can be localized within the muscle movement system, the only additional system that becomes engaged during overt speech.

The Number of Muscle Movement Components

If stuttering represents a muscle movement disorder, then decreasing the number of muscle movement nodes that become activated during speech should decrease the probability of stuttering. A study by Brenner et al. (1972) strongly supported this prediction. They had stutterers produce sentences in three different ways: mouthing (moving only the lips), whispering, and full-fledged articulation. The results showed that severity of stuttering varied with the number of muscles involved. The stutterers stuttered least when moving only their lips, more when whispering, and most when engaging in full-fledged articulation.

Muscle Movement Problems

If stuttering reflects a muscle movement defect, then stutterers may exhibit this defect independently of the phonological processes that normally control movements of the speech musculature. This is exactly what studies by McFarlane and Prins (1978) and Cross and Luper (1979) showed. They compared reaction times of stutterers and nonstutterers using speech muscles not to make speech sounds but simply to react as quickly as possible, for example, closing the lips in response to a pure tone stimulus. Stutterers were slower at initiating these movements than were nonstutterers, a finding that cannot be explained at the phonological level. A follow-up study by Reich, Till, and Goldsmith (1981) showed that these slow reaction times are confined to the muscles normally used for speech. Stutterers and nonstutterers exhibited equivalent reaction times when using nonspeech muscles, for example, when pressing a key in response to a tone.

Absence of Corresponding Perceptual Deficit

Stutterers show no corresponding deficits in the perception of speech sounds. They never misperceive someone to say "p-p-p-please," when in fact the person said "please." This lack of perceptual deficit fits the hypothesis that stuttering originates in muscle movement nodes (which are specific to production), rather than in sensory analysis nodes (which are specific to perception), or in phonological nodes (which play a role in both perception and production).

The Time Characteristics and Automaticity of Stuttering

Two salient characteristics distinguish muscle movement nodes from higher level nodes: automaticity and rapid activation. As D. G. MacKay (1982) points out, muscle movement nodes receive much more practice than higher level nodes and are much more likely to achieve automaticity, including independence from conscious control. Muscle movement nodes are also activated much more rapidly than higher level nodes. The lowest level muscle movement nodes become activated within a time frame of milliseconds, segments nodes within a time frame of tens of milliseconds, and lexical nodes within a time frame of hundreds of milliseconds.

Both of these characteristics (automaticity and rapid activation) are also characteristics of stuttering, as would be expected under the hypothesis that stuttering represents a muscle movement disorder. Stuttering is automatic and beyond conscious control. Stutterers are unable to alter the way they execute muscle movements so as to avoid stuttering, just as normal speakers are unable to select or alter their muscle movements so as to produce a contextually inappropriate allophone. Stuttering misarticulations also occupy a time frame characteristic of muscle movement nodes rather than phonological nodes. Like muscle movements, stuttering repetitions occur within milliseconds. Durations of 20 ms or less are not atypical for repeated lip movements in stuttering. And stutterers sometimes exhibit movement asynchronies (e.g., between lip and jaw movements) that are so brief as to be inaudible to the unaided ear (Zimmermann, 1980).

The Node Structure Theory of Intrinsic Stuttering

Can the node structure theory provide an integrated account of stuttering and other speech errors? I begin with a review of the basis for other errors in the node structure theory and then examine whether repetitions, prolongations, and blocks can be explained within the same framework.

Error-free output occurs under the theory when an "intended-to-be-activated" content node accumulates greater priming than any other node in its domain and becomes activated. The intended-to-be-activated node is the one that is receiving priming from a superordinate node in the output sequence, that is, the directly connected content node immediately higher in the hierarchy. This priming summates over time and eventually exceeds the priming of all other nodes in the domain, say, by time t_2 in Figure 8.7. At or after this point in time, the triggering mechanism will activate the intended-to-be-activated node under the most-primed-wins principle, and the output is error free.

Consider now substitution errors, where one component substitutes another, as in the anticipation error *coat cutting* instead of *throat cutting*. Substitution errors occur whenever an intended-to-be-activated node acquires less priming than some other node in its domain when the triggering mechanism is applied. Thus, the fundamental cause of substitution errors is that other, "extraneous" nodes in a domain are also receiving priming, and this extraneous priming or noise sometimes exceeds the systematically increasing priming for the intended-to-be-activated node at the time when the triggering mechanism is applied. As a

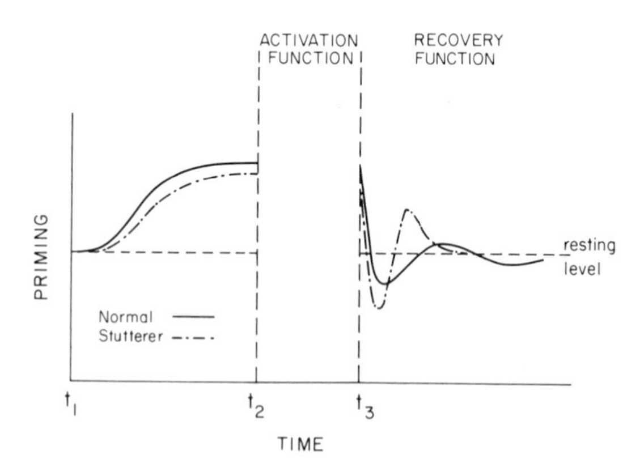

FIGURE 8.7. The priming function (relating degree of priming and time following onset of priming, t_1), activation function (following application of the triggering mechanism at t_2), and recovery function following onset of self-inhibition at t_3) of stutterers and nonstutterers. See text for explanation. (From "Stuttering as a Sequencing and Timing Disorder" by D. G. MacKay & M. MacDonald in *Nature and Treatment of Stuttering: New Directions* (p. 273) edited by W. H. Perkins & R. Curlee, 1984, San Diego: College-Hill. Copyright 1984 by College-Hill Press. Reprinted by permission.)

consequence, the wrong node becomes activated under the most-primed-wins principle, and an error occurs. Because of the way priming summates for every node in a hierarchy (Figure 8.7), substitution errors will be more likely at faster rates of speech. At faster rates the triggering mechanism is applied sooner following onset of priming (t_1), allowing less buildup of priming (D. G. MacKay, 1982).

Now consider repetitions, prolongations, and blocks within this framework for explaining substitution errors. During repetitions, a just-activated component becomes reactivated, and the issue is how the same sequence and content nodes can acquire enough priming following activation to become most primed in their domain and thereby reactivated. One possibility is that muscle movement nodes of stutterers may exhibit an abnormal priming and recovery cycle (illustrated in Figure 8.7), such that priming builds up abnormally slowly and rebounds abnormaly sharply following self-inhibition (as would occur if inhibitory satellites and parent neurons alike had abnormally high thresholds or insensitivity to cross-connection priming). As a consequence, just-activated sequence nodes have a high probability of again becoming most primed in their domain and reactivated with the next pulse from the timing node. Because a just-activated content node also rebounds sharply following self-inhibition, a repetition such a *p-practice* will result.

This reactivation cycle can of course reiterate, so that the same movement is repeated three or four times, but the reactivation cycle cannot go on indefinitely. Because satiation increases with repeated activation, a repeatedly reactivated node cannot rise to such high levels of priming on rebound from inhibition, and so eventually fails to achieve more priming than the next-to-be-activated node. The fact that priming for the next-to-be-activated nodes continues to summate during stuttering also mitigates against long strings of repetitions. Under the theory, the longer the period of stuttering, the greater the likelihood that the next pulse from the triggering mechanism will activate the correct node.

The same slow buildup of priming that contributes to repetitions under the theory can also cause prolongations and blocks. During a prolongation, the articulators become locked into a continuant (open) position, while during a block, they become locked into an obstruent (closed) position, preventing speech by prohibiting airflow. Again, the reason is that in stutterers, inhibitory satellites of muscle movement nodes accumulate priming abnormally slowly, causing delays in self-inhibition of the parent node. The muscle movement nodes of stutterers can therefore remain activated for longer than normal durations, as occurs during prolongations and blocks.

Major disfluencies occur under this account when several muscle movement nodes happen to malfunction as a group. However, minor disfluencies, inaudible to the human ear, may occur when only a few muscle movement nodes malfunction. This may explain why cineradiographic analyses of stutterers' utterances exhibit abnormal transitions between sounds and asynchronies between lip and jaw movements, even though the utterances sound fluent to the unaided ear (Zimmermann, 1980); some fraction of the underlying muscle movement nodes are being activated at the wrong times.

Stuttering, Sequencing, and Timing

As discussed so far, stuttering reflects a disruption in the ability to produce the next muscle movement component in sequence, a problem involving content and sequence nodes. Several sources of evidence support this postulated sequencing difficulty (Van Riper, 1982), but stutterers also have a timing problem that is relevant to the node structure theory. As already noted, stutterers exhibit asynchronies between the muscles for articulation, respiration, and voicing, not only during overt instances of stuttering but also during seemingly fluent speech (Zimmermann, 1980). Relative to nonstutterers, stutterers also have difficulty duplicating the timing of a sequence of finger taps (M. H. Cooper & Allen, 1977), as if the timing nodes governing nonspeech activities are also more variable and susceptible to mistiming. Perhaps timing nodes suffer the same slow buildup of priming and sharp rebound from self-inhibition as content and sequence nodes. The observed deviations from periodicity may therefore occur when the sharp rebound suffices to reactivate the timing node.

The possibility that stuttering reflects a periodic disruption in the timing of muscle movements may explain phenomena such as the "rhythm effect," where fluency is enhanced when a stutterer speaks in time with a metronome or any other rhythmic stimulus (visual, auditory, or tactile), provided that the rhythm is not abnormally fast (Van Riper, 1982). That is, an externally generated rhythm to a shared perception–production timing node (Chapter 5) may help facilitate the timing of motor patterns that are prone to asynchrony in stutterers. Timing factors may also explain why stutterers often become more fluent when singing (Van Riper, 1982). The externally provided rhythm of the notes may help establish periodicity for the timing nodes controlling syllable production (although speed–accuracy trade-off explanations remain to be ruled out).

Practical Implications

The fact that stuttering originates within the muscle movement system in the node structure theory suggests that manipulating muscle movement factors can directly ameliorate stuttering. However, muscle movement factors are difficult to manipulate. Processes within the muscle movement system are largely automatic, that is, overlearned, fast, unconscious, beyond voluntary control, and in general unmodifiable (D. G. MacKay, 1982; 1987).

Muscle movement processes are also extremely complex. For example, over 100 different muscles may become engaged in producing a single word, and each muscle must get its appropriate nervous impulses at the required moment and in the proper sequence if the word is to be spoken without disruption. Because we currently don't know how muscle movement nodes interconnect and interact in real time, let alone how to modify these interactions, fundamental solutions to the problem of stuttering are a long way off. Understanding stuttering and the structure and dynamics of the muscle movement system for speech will remain an important theoretical goal for many decades to come.

9
Perceptual Feedback in the Detection and Correction of Errors

> Self-repairs are . . . rather complex phenomena . . . they involve quite disparate phonetic processes, such as self-monitoring, the production and detection of phonetic, lexical and other types of speech errors, self-interruption, prosodic marking of the connection.
>
> (Levelt, 1984, p. 105)

Self-inhibition is a central concept in the node structure theory. In addition to explaining the many sources of evidence discussed in Chapter 8, self-inhibition is needed to explain how perceptual feedback is processed during an ongoing action and to explain how errors are detected and corrected. Self-inhibition also makes sense of the way speech production becomes disrupted when normal speakers hear the sound of their own voice amplified and delayed by about 0.2 s. Self-inhibition even contributes to "node commitment," as I call the process of forming new or functional connections between nodes. However, I have developed the last two topics elsewhere (Chapter 10, and MacKay, 1987). The present chapter discusses only feedback processing and the detection and correction of errors. I first examine how errors are detected and corrected and the constraints these phenomena place on theories of the relation between perception and action. I then take these constraints into account in developing a theory of mechanisms underlying error detection and correction and the processing of self-produced feedback in general.

Constraints on Theories of Error Detection

Perceptual feedback plays a major role in detecting self-produced errors, and theories of feedback processing must explain (at least) four general characteristics of error detection: the rapidity of error detection, the detection of internal errors, differences between detecting correct versus incorrect responses, and differences between detecting self-produced versus other-produced errors.

The Time Characteristics of Self-Interruption

Studies of spontaneous self-corrections, reviewed in Levelt (1984), indicate that errors can be detected very quickly, even before the error has been completed in the surface output. After making an error in word selection, subjects often interrupt themselves immediately, sometimes before they finish the word containing the error. For example, a subject in Levelt (1984) began to say the word *black* but stopped after the initial /b/ and then produced the intended word *white*, as in "*b-* . . . er, *white*." These within-word interruptions usually occur when the output is factually incorrect (as in this example) and not simply infelicitious or open to misinterpretation. Speakers not only detect and correct their errors very rapidly, but they apparently comprehend the factual versus infelicitious nature of their error before they interrupt and correct themselves.

Internal Error Detection

Although the evidence previously discussed suggests that a speaker can *sometimes* detect errors before their full-blown appearance in the surface output, stronger evidence is needed on this point. After all, corrections sometimes fail to occur, or they may follow a full-blown error with considerable lag. Experiments on the detection of errors in internal speech provide the additional support, indicating that under appropriate circumstances, *all* (phonological) errors can be detected internally, without ever appearing in the surface output. Dell (1980) showed that when subjects produce tongue twisters, either overtly or to themselves, they make as many errors and detect the occurrence of these errors as often during internal speech as during overt speech (Chapter 8). Theories of error detection must therefore explain how errors can be detected before their appearance in the surface output, and why *external* feedback is unnecessary for error detection.

Self-Produced Versus Other-Produced Error Detection

Error detection differs in interesting ways, depending on whether the error is self-produced or other-produced. Other-produced errors are more easily detected for units that are large and meaningful, such as words, rather than small and meaningless, such as phonemes and phoneme clusters (Tent & Clark, 1980). Self-produced errors, on the other hand, are detected and corrected about equally often, regardless of the size or meaningfulness of the units involved (Nooteboom, 1980). Apparently speakers respond to their own errors with equal sensitivity, regardless of error size or type, whereas listeners are especially sensitive to errors involving large and meaningful units (Cutler, 1982), a fundamental constraint on theories of error detection.

Detection of Correct Versus Incorrect Responses

Detecting that a response is correct takes more time than detecting that a response is in error. Rabbitt, Vyas, and Feamley (1975) examined how rapidly subjects performing a two-choice reaction time task could indicate that a given response was either correct or incorrect, and they found that these "validation times" were consistently longer for correct responses than for incorrect responses.

Editor Theories of Error Detection

Current theories have encountered difficulties in explaining the nature of error detection. To illustrate some of these difficulties, I examine the process of error detection in the class of theories known as "editor theories." I then develop a new and somewhat simpler theory of error detection that seems to overcome these problems.

The defining characteristic of an editor theory is a mechanism that "listens to" the visual, auditory, kinesthetic, or other feedback that results from output; compares this feedback with the intended output; identifies errors; and then computes corrections using a duplicate copy of information originally available to the motor control system. Baars et al. (1975) and Keele (1986) provide recent examples that fall within this class, but the editor theory tradition extends back at least to Freud (1901/1914). Recent variants in the editor theory class have added the concept of efference-copy or "feedforward," a preliminary representation of an about-to-be-produced action that enables editing before the actual occurrence of the output (Keele, 1986).

Problems with even the latest editor theories are everywhere apparent. If an editor knows the correct output all along, one wonders why the correct output wasn't generated in the first place. One also wonders why errors sometimes pass uncorrected. According to data of Meringer (1908) and Meringer and Mayer (1895), speakers only correct about 60% of all word substitution errors, as in "*table*, er, I mean *chair*." Perhaps the editor is subject to the same laws of fatigue or haste as the output system and therefore tends to err at the very time that the output system makes an error. However, this explanation calls into question the purpose of an editor that tends to break down at the time when it is needed.

Another set of problems arises from the assumption, implicit in editor theories, that perception and production involve separate components at all levels. That is, editors use the same mechanisms for the perception of errors as for perception in general. Then, once an error has been detected, the editor computes corrections for execution by a production system that is logically distinct and separate from the perceptual system. However, if systems for perception and production are distinct and separate, then the many parallels and interactions between perception and production discussed throughout this book become difficult to explain.

Editor theories also have difficulty explaining asymmetries between detection of self-produced versus other-produced errors. If an editor uses general perceptual mechanisms to detect errors, then different types of errors should be equally easy to detect in self-produced and other-produced speech. As already noted, however, experimental data do not support this prediction. Lexical substitution errors (e.g., *table* misproduced as *chair*) are detected more frequently than phonological substitution errors (e.g., *publication* misproduced as *tublication*) in other-produced speech (Tent & Clark, 1980) but equally frequently in self-produced speech (Nooteboom, 1980).

The fact that detecting correct responses takes more time than detecting incorrect responses (Rabbitt et al., 1975) presents another problem for editor theories. As Broadbent (1971) points out, if the output is continually monitored for correctness versus incorrectness, editor theories predict the opposite result. Because correct responses occur more often than errors, the editor should be more practiced, and therefore faster at monitoring correct responses. The failure of this and other predictions suggests that editor theories require modification along lines discussed in the following section.

The Node Structure Theory of Error Detection

The node structure theory can be considered to combine aspects of editor theories and corollary discharge theories (discussed later in this chapter) to provide a coherent account of mechanisms underlying feedback processing and the detection of everyday errors. I begin with an analysis of the relationship between top-down and bottom-up (feedback) processes involving mental nodes and then develop an account of the detection of errors in self-produced versus other-produced speech, the time characteristics of error detection, and the time required to detect correct versus incorrect responses.

The Role of Internal Feedback in Error Detection

Both internal and external feedback can play a role in the detection of errors under the node structure theory. Recall that internal feedback consists of bottom-up priming transmitted from a subordinate node to its superordinate node as soon as the subordinate node becomes activated during production. External feedback likewise primes a just-activated superordinate node bottom-up, but this priming arrives later, following sensory analysis of the auditory or other perceptual consequences of the action.

The role of internal feedback in the detection of self-produced errors is extremely simple in the node structure theory but depends on understanding how errors occur. As discussed in D. G. MacKay (1982; and Chapter 7), output errors occur when a primed-from-above node fails to achieve greatest priming in its domain, and some other node becomes activated. An example is the error "Put the bag in the . . . ." instead of "Put the box in the car" (from Goldstein, 1968). Here the intended or primed-from-above node is *box*(noun), but *bag*(noun) has acquired greatest priming in the (noun) domain and has become activated under the most-primed-wins principle, giving rise to the error *bag*, rather than the intended *box*.

Such an error can be perceived, although not necessarily prevented, even before it appears in the surface output. Because activation is an inherent aspect of both perception and production, and because perception and production use the same mental nodes, activating the wrong node in production can cause virtually immediate perception of the error. In the preceding example, erroneous activation of *bag*(noun) implies potentially immediate perception of the unintended concept, "bag." This explains how errors can be detected so quickly. Perception of an error begins in the node structure theory even before the error has been expressed. Of course, as in perception in general, activation is necessary, but not sufficient for *awareness* of an error. Moreover, error detection is necessary, but not sufficient for error *correction*. Because *listeners* often fail to detect errors (e.g., Marslen-Wilson & Tyler, 1980), *speakers* can adopt a fairly liberal criterion in deciding whether or not to correct their own errors.

This view predicts that awareness of a specific error should always be part of awareness that an error has occurred. Because *perceiving* an error is yoked to activating the mental node that *produces* the error in the first place, speakers will never perceive that they made some error, but not know what the error was. Perception of self-produced error will never become dissociated from the specific content of the error that has occurred. Interestingly, this will *not* be true for perception of "other-produced" errors under the theory. For other-produced errors, the perceiver *can* become aware that an error has occurred, but not know with certainty what the error is. In what follows, I discuss various other predictions derived from the node structure theory of error detection and show how the theory handles the constraints on theories of error detection discussed previously.

Detection of Internal Errors

Although external feedback can facilitate error detection under the node structure theory, internal feedback is sufficient for detecting errors involving mental nodes. This explains why phonological errors can be detected in internal speech (Dell, 1980) without actually occurring in the surface output. An error can be detected as soon as the wrong node in an output hierarchy has been activated, that is, before and independently from activation of the lower level nodes for expressing the error and for processing the external feedback that results.

Detection of Correct Versus Incorrect Responses

Under the node structure theory, detecting an error involves a simpler and more direct process than detecting a correct response. The reason is that internal feedback (bottom-up priming) can lead directly to perception in the case of errors, but internal feedback is canceled by self-inhibition in the case of correct responses.

Errors involve the activation of an extraneous node, which results in bottom-up convergent priming of higher level nodes. These higher level nodes are not undergoing self-inhibition and can therefore become activated to provide a basis for detecting the error. Indeed, higher level nodes receiving bottom-up priming as a result of an error are extremely likely to become activated under the most-primed-wins principle, relative to the nodes for producing the remainder of the output sequence. These intended-to-be-activated nodes are only receiving top-down priming, which as discussed in Chapter 7, is nonconvergent and therefore weak, relative to the bottom-up priming from an error, which converges and summates and is therefore relatively strong.

The same bottom-up convergent priming also occurs when the appropriate nodes have been activated in producing the correct or intended output. The difference is that higher level nodes representing a correct output *are* undergoing self-inhibition and so cannot become reactivated and provide the basis for detecting that the output is in fact error free. In the case of a correct response, self-inhibition cancels the internal feedback, which provides the normal signal for error detection. As a result, verifying that a response is correct will take more time than detecting that a response is in error, as Rabbitt et al. (1975) observed.

Self-Produced Versus Other-Produced Error Detection

The node structure theory suggests an interesting explanation for differences between the perception of self-produced versus other-produced errors. An output error will occur when a single extraneous node in any domain in any system becomes activated and activation of a node enables perception, no matter what the size of the surface units involved (segment, segment clusters, syllables, words, or phrases). This means that speakers will perceive self-produced errors equally often for small versus large units (all other factors being equal), exactly as Nooteboom (1980) observed.

However, perception of *other-produced* errors is quite different. Because of the principle of higher level activation, other-produced errors involving (smaller) phonological or phonetic units are likely to pass undetected. Because listeners don't normally activate either phonetic or phonological nodes during everyday perception of other-produced speech, listeners are less likely to perceive phonological and phonetic errors, just as Tent and Clark (1980) observed. However, phonological errors become easier to perceive if these errors alter the higher level interpretation of an utterance, because the nodes representing higher level interpretations routinely become activated under the principle of higher level activation.

The node structure theory also predicts that for normal inputs occurring at normal rates, listeners will be able to detect other-produced errors either at the word level or at lower (phoneme or letter) levels but not at both levels simultaneously (D. G. MacKay, 1987). Although this prediction remains to be tested, the nature of proofreading provides tentative or preliminary support. Misspellings are notoriously difficult to detect when reading for higher level content, and understanding the content of a passage is notoriously difficult when proofreading for misspellings.

A Comparison of Node Structure and Editor Theories

The node structure theory of error detection introduces a series of refinements that overcome the problems of editor theories discussed earlier. In brief overview–summary, the node structure theory resembles editor theories in the use of identical mechanisms for perceiving self-produced and other-produced inputs. However, because lower level (e.g., phonological) nodes become activated and self-inhibited during *production*, but not during *perception*, the node structure theory predicts major differences between detection of self-produced versus other-produced phonological errors.

The error detection mechanism in the node structure theory consists of many distributed processors, instead of a single centralized processor (such as the editor in editor theories). The distributed processors in the node structure theory correspond to the nodes that become primed and activated when producing the original output. The node structure theory also detects errors automatically and directly instead of "indirectly" via a comparison or matching process, as in editor theories. On the other hand, the node structure theory cannot immediately and directly perceive the correctness of a correctly produced output (unlike editor theories) because of the self-inhibition that follows activation of a correct or primed-from-above mental node.

Other Aspects of Feedback Processing

Theories of feedback processing must explain more than just error detection, and in this section I examine two other constraints on a general account of feedback processing: the missing feedback effect and the role of feedback in learning.

The Missing Feedback Effect

Chapter 4 described the verbal transformation effect, the fact that perception changes when an acoustically presented word is repeated via tape loop for prolonged periods. After many repetitions of the word *pace*, for example, subjects reported hearing words such as *face*, *paste*, *base*, *taste*, or *case*, and the rate of perceptual change increased as a function of time or number of repetitions (Warren, 1968). What must be explained in theories of perceptual feedback is why similar perceptual changes *fail to occur* when subjects produce the repeated input *themselves*. Lackner (1974) showed that the auditory feedback that accompanies repeated *production* of a word fails to trigger verbal transformations. Lackner's subjects repeated a word every 500 ms for several minutes, and they later listened to a tape recording of their own output. The subjects experienced the usual transformations when *listening* to the tape repeated word, but for some reason, experienced no perceptual transformations when *producing* the repeated word. This missing feedback effect is interesting, because acoustic events at the ear are identical when hearing the input during versus after production.

Why does on-line auditory feedback fail to trigger verbal transformations? Lackner (1974) and Warren (1968) attributed the missing feedback effect to corollary discharge, an efference-copy that accompanies the motor command to produce a word. This corollary discharge cancels or inhibits the external (proprioceptive and auditory) feedback resulting from producing the word, so that the on-line auditory input fails to bring about the fatigue-induced perceptual changes that are typically observed in verbal transformation experiments.

Taken by itself, the corollary discharge hypothesis has difficulty explaining the many interactions between speech perception and production discussed throughout this book (especially the effects of delayed auditory feedback, Chapter 10). Corollary discharge also fails to explain an additional symmetry in Lackner's own (1974) data, namely that no *production errors* resembling the perceptual errors occurred while subjects actively repeated the words. And Bridgeman (1986), Steinbach (1985), and D. M. MacKay (1973) summarize several additional problems with corollary discharge as an explanation of other aspects of perception, such as the phenomenon of perceptual stability following voluntary eye movements and the occurrence of visual suppression during saccades (the fact that psychophysical thresholds increase by about a half-log unit when faint test flashes are presented during or just before a saccade).

Corollary Discharge as Self-Inhibition

Under the node structure theory, the self-inhibition that follows activation of mental nodes is responsible for many of the phenomena that have been attributed to corollary discharge, and the missing feedback effect provides a good illustration of this corollary discharge as self-inhibition hypothesis. If corollary discharge corresponds to the activation of inhibitory satellites, the missing feedback effect becomes immediately apparent. Mental nodes for producing and perceiving a word such as *police* are identical, so that when someone repeats a word such as *police*, auditory feedback returns as priming to the just-activated nodes that produced it, but has little effect, because these nodes are undergoing self-inhibition following production. Self-inhibition of lower level nodes also prevents returning external feedback from reaching higher level nodes, so that self-produced inputs fail to cause verbal transformations, which is the missing feedback effect.

The corollary discharge as self-inhibition hypothesis also explains the absence of output errors during repeated *production* of a word. For example, repeating the word *police* causes satiation of the corresponding mental nodes, but because top-down priming is unique (i.e., only a single node normally receives first-order priming in any given domain at any given time), only *police*(noun) and no other lexical node receives systematically increasing top-down priming and becomes activated under the most-primed-wins principle. The uniqueness of top-down priming during production therefore reduces the likelihood of *production* errors resembling those that occur in perception and explains this additional asymmetry in Lackner's (1974) data.

The node structure characterization of the missing feedback effect makes several interesting sets of predictions not made by other theories. One concerns the effects of production on perception. If subjects produce a word repeatedly for, say, 10 minutes, either *internally or overtly*, and then listen to the same word repeated via tape loop, the subsequent rate of perceptual change will be *at least as great* as if they had already heard the word repeated for 10 minutes. Another set of predictions concerns a *production* condition where subjects repeat a word (again either internally or overtly) while hearing a *phonologically and semantically different* input word repeated over earphones at the same time. After, say, 10 minutes, the subjects are signaled to stop production and report perceptual transformations of the input word as it continues to repeat for, say, 5 minutes. Under the node structure theory, perceptual transformations of the input word will be significantly slower in this *production* condition than in the usual *listening* condition where subjects attend to the repeating input word for the entire 15-minute period.

Feedback and the Formation of New Connections

Internal and external feedback have other functions in addition to error correction in the node structure theory. One concerns the commitment of new connections, a process discussed in detail elsewhere (D. G. MacKay, 1987). What follows is a severely truncated summary.

Feedback is required during early stages of acquiring a skill in order to ensure that an action is appropriate or has the desired effect. In learning to produce a German trilled /r/, for example, it is essential to hear the trilled /r/ that one produces in order to know whether one's muscle movements achieved the desired result. Without external feedback during this early phase of acquisition, practice can strengthen the wrong connections and lead to inappropriate actions (D. G. MacKay, 1981). During later stages of skill acquisition, the linkage strength of already formed connections can become further strengthened in the complete absence of feedback, and this explains how mental practice can improve skilled performance without the help of either sensory or experimenter-generated feedback (D. G. MacKay, 1981).

In short, external feedback is unnecessary for execution of behavior during later stages of skill acquisition, when connections for an output hierarchy have already been formed and strengthened. This explains why experimentally deafferented monkeys can accurately produce a previously learned and highly practiced response (Bizzi, Dev, Morasso, & Polit, 1978). It also explains why adults can in general produce intelligible speech when their auditory feedback has been masked, distorted, or eliminated altogether as a result of acquired deafness (Siegel & Pick, 1974). The effects of delayed and amplified auditory feedback (DAAF) on speech production provide the only alleged counterexample of feedback seeming to influence the control of highly skilled behavior. When auditory feedback is amplified and delayed by about 0.2 s, normal speech production becomes severely disrupted. This remarkable and highly reliable phenomenon is discussed in detail in Chapter 10, and contrary to surface appearances, it does not support the feedback control hypothesis.

Constraints on Theories of Error Correction

Once an error has been detected, how is it corrected? Error correction does not proceed on a trial-and-error basis. Almost immediately after becoming aware of making an error, we also become aware of the exact locus and nature of the error, how to correct it, and how to signal the nature of the error to the listener. Theories of error correction must therefore explain three general classes of phenomena: the process of error signaling, the time course of error correction, and regularities in the way that errors are corrected, such as the structural and lexical identity effects discussed in a following section.

The Time Course of Error Correction

The process of error correction seems to begin almost immediately after an error has been committed. After making an error in a multiple choice reaction time task for example, subjects do not go through the entire process of deciding all over again what response to make. Rabbitt and Phillips (1967) showed that the interval between an error and its correction is much shorter than the average time required to produce the original response correctly. Moreover, the time required to detect and correct an error seems to remain constant when the original response time is extended by reducing stimulus–response compatibility, increasing the number of response alternatives, or increasing the number of stimuli appropriate to each response. As Broadbent (1971, p. 305) points out, the fact that task difficulty fails to influence the time to correct errors rules out the hypothesis that errors are corrected by reiterating the original response process, even if the correction process can begin before the occurrence of the error in the surface output. As Broadbent (1971) also points out, the rapidity of error detection and correction is surprising in view of the refractory effects that are generally observed when subjects must make two responses in rapid succession (Chapter 8). The (hypothesized) internally generated "error signal," which is used for correcting an error in editor theories, is clearly unlike an external signal, which follows soon after the first in the psychological refractory paradigm.

Error Signaling

When speakers stop to correct an error during speech production, they often introduce a term such as "uh," "er," or "I mean," which signals both the occurrence and nature of the error. These error signals fall into two classes: rejection error signals and supportive error signals. Speakers usually introduce rejection error signals when the error results in a *factually incorrect* statement, as in "feeding it into the computer, *sorry* [*no, rather*], the printer." A change in prosody often accompanies these rejection signals, so that the repair receives, for example, greater stress than the trouble segment. However, when the error results in a *factually correct* but inadventitious statement, prosody remains normal, and a supportive error signal is introduced, as in, "Go into the room, *thus* [*that is*], the kitchen" (Levelt & Cutler, 1983). Theories of error correction must explain how speakers can so rapidly distinguish and signal these two types of errors.

The Structural and Lexical Identity Effects

Words used in correction usually maintain the same syntactic structure as the words they correct, as if the correction and corrected words represent conjoints in a coordinate structure (e.g., nouns joined by *and*). In "Is he seeing, er, interviewing Mary?" for example, the erroneous word, *seeing*, belongs to the same syntactic class as the correction, *interviewing* (Levelt, 1984).

Indeed, erroneous words and their corrections are often more than just structurally identical. Corrections that repeat ideas in the original utterance usually copy the original words verbatim, and this is especially true of factual errors, as in, "He talked frequently with his sister, uh, I mean, he talked frequently with his *mother* [emphasized]." Theories of error correction must therefore explain why corrections are usually structurally *and* (if possible) lexically identical to the originally intended words.

The Node Structure Theory of Error Correction

The efficiency of error correction is readily explained in the node structure theory. As previously discussed, error detection occurs almost as soon as the error itself in the theory, and error correction can begin shortly thereafter, when the correct or primed-from-above node becomes activated. By way of illustration, consider again the error "Put the *bag*, I mean, *box* in the car." Assume that *bag*(noun) has been activated, that the error has been detected, and that the flow of speech has been interrupted. During this time, the appropriate or primed-from-above node, *box*(noun), continues to accumulate priming from its immediately superordinate noun phrase node (until self-inhibition begins), as well as from other nodes representing, say, the situational context. This means that if the most recently activated sequence node (NOUN) is reactivated, the appropriate node, *box*(noun) in this example, will be activated, resulting in direct perception of the intended output. Perception of the intended output in turn enables immediate inferences as to the nature of the error and the type of error signal required.

Because the appropriate node has now been activated, the appropriate output can also be produced. Indeed, activating the appropriate or primed-from-above node is *part of* producing the correct response. No special mechanism and no extra time is required for computing which elements are in error or how to correct them. The correct response is immediately available under the node structure theory. The correction is produced by activating the same sequence nodes and hierarchy of content nodes, except of course, for the one content node that was activated in error (plus all of its subordinate nodes). At the time when repair occurs, the appropriate node in the domain of nodes containing the error will have greatest priming and become activated as part of the repair.

Error-Correction Errors

Besides being consistent with current constraints on theories of error correction, the node structure theory makes some interesting predictions for future test. One concerns a class of "error-correction" errors. The node structure mechanism for error correction is itself susceptible to errors. The main prerequisite for error-correction errors is that during the normal course of producing the remainder of a sentence, the sequence node responsible for an error is due to become activated again soon after the error occurs.

By way of illustration, consider again the substitution error "Put the bag in the . . ." instead of "Put the box in the car." In producing the intended sentence, the NOUN sequence node must become activated in quick succession, first to produce *box* and then to produce *car*. Thus, if the rate of speech is sufficiently rapid when the substitution of *bag* for *box* occurs, the theory predicts a double error of the form, "Put the bag in the box." That is, the correct noun for the earlier slot will become activated (now erroneously) in the subsequent noun slot. The reason is this: After the inappropriate node *bag*(noun) has been activated, the originally appropriate node, *box*(noun), continues to accumulate priming, is likely to achieve greatest priming in the (*noun*) domain, and becomes activated automatically when its sequence node *NOUN* is applied to the (noun) domain soon after in the course of producing the remainder of the sentence. It is as if a noun has been corrected but the wrong noun!

Because error detection and self-interruption is so rapid, however, the likelihood of observing such error-correction errors is relatively low. Nevertheless, error-correction errors do occur (see Goldstein, 1968 for examples), and their occurrence can be viewed as support for the node structure theory of error correction. (Related processes are also required in explaining the subtly different "bumper-car errors" in Stemberger, 1985.)

The Structure of Error Correction

The node structure theory readily explains the structural and lexical identity effects, the fact that corrections copy the syntactic structure and, if necessary, as many correct words as possible from the original utterance. Under the node structure theory, corrections occur by simply reactiving the same sequence node(s) following detection of an error, and in the case of correct or "copied" words, this means activating the same content nodes as well. Only the content node that was activated in error will not be reactivated. This "erroneous" node will be recovering from activation, and the appropriate node will (because of temporal summation) achieve greatest priming in its domain and become activated automatically as part of the repair. Under the theory, this appropriate node typically will constitute the only difference between the correction and the corrected output, error signals aside. Thus, even the prosody of the new or corrected utterance should correspond to that of the original or intended utterance, and this is exactly what Levelt (1984) found. Levelt (1984) recorded a large number of naturally occurring errors, erased any error signals or repeated words, and spliced the correction itself onto the original utterance. He then had subjects listen to the resulting utterances, and found that the spliced utterances sounded completely natural, even when up to 3 s of output had been deleted.

10
Disruptive Effects of Feedback

All purposeful behavior may be considered to require negative feedback.
(Rosenblueth, Wiener, & Bigelow, 1943, pp. 22).

This is an exciting period for the study of action.
(Gentner, 1985, p. 184).

Although feedback can help in detecting and correcting errors, and in learning new behaviors, feedback can also disrupt ongoing action. Many perceptual–motor systems exhibit feedback-induced disruptions, and speech production under conditions of "delayed auditory feedback" provides the most dramatic and carefully studied example. When auditory feedback from speech is recorded and then played back with amplification to the ears after a delay of about 0.2 s, speech becomes severely disrupted. Under these conditions, proficient speakers repeat, prolong, and substitute speech sounds, sometimes producing phonemes that are not part of any language familiar to them (B. S. Lee, 1950). The present chapter examines this and other feedback-induced disruptions and the constraints they impose on theoretical relationships between perception and action.

Experiments on delayed and amplified auditory feedback (DAAF) flourished in the 1950s and 1960s, because the phenomenon seemed compatible with the cybernetic or feedback control theory in vogue at the time (Neumann, 1984). However, the last 15 years have seen virtually no studies of DAAF, not because we now understand the phenomenon or because it proved unimportant or unreplicable; the effects of DAAF are highly reliable, and remain central to an understanding of the relation between perception and action. Rather, the problem was that the theory that originally stimulated interest in the phenomenon proved inadequate for explaining the new experiments conducted in its name. Only a new theory with predictions for future test can revive interest in DAAF, and my goal here is to develop such a theory. In the Epilogue, which follows this chapter, I examine some additional lessons of DAAF for the many other "abandoned phenomena" in psychology.

Constraints on Explanations of Feedback Disruption

Theories of feedback processing must account for five basic characteristics of DAAF disruption: the conditions for disruption, the automaticity and generality of the phenomenon, factors influencing the delay that produces maximal interference, and factors influencing the degree of disruption.

Basic Conditions for DAAF Disruption

The most basic condition for disruption is, of course, the fraction of a second delay in external feedback. Another basic condition, frequently overlooked in explanations of feedback disruption, is amplification of the returning feedback. Articulatory errors do not occur unless auditory feedback returns at louder than normal levels (see, for example, Fairbanks & Guttman, 1958). The importance of amplification prompted my use of the term *DAAF* (delayed and amplified auditory feedback), rather than the usual *DAF* (delayed auditory feedback).

At one time, amplification was assumed to have a masking function. Unless masked by *amplified* auditory input, bone-conducted feedback could be used for feedback control (because the skull transmits voice-correlated vibrations that cannot be delayed). However, masking can't be the sole effect of amplification (see also Howell & Powell, 1984). DAAF errors continue to occur when low-frequency auditory components are eliminated by means of a low-band pass filter (Hull, 1952), whereas under the masking hypothesis, low-band pass filters should fully restore fluency: lower level frequencies that bone-conduction primarily transmits will no longer be masked. Moreover, increasing feedback amplitude over a wide range continues to increase the probability of errors, long after air-conducted feedback could be assumed to have masked the bone-conducted feedback (Black, 1951). Some other factor is therefore needed to explain the effect of feedback amplification on DAAF errors (see also Howell & Archer, 1984).

Automaticity of the Phenomenon

As many investigators have noted, the automaticity of DAAF interference resembles pathological stuttering. Like pathological stuttering, DAAF errors cannot be voluntarily corrected or avoided if subjects are speaking at maximal rate and without pauses. Also like pathological stuttering, practice with disrupted output fails to eliminate DAAF interference (unless subjects are permitted to speak in short bursts, thereby desynchronizing their output and returning feedback; see Smith, 1962; see also D. G. MacKay & Bowman, 1969, for a transfer paradigm where practice *can* genuinely reduce DAAF interference).

These similarities aside, pathological and DAAF stuttering differ in some respects that also require theoretical explanation. One difference concerns the loudness of speech. Speakers hearing DAAF involuntarily speak with greater than normal amplitude and increase their output amplitude in proportion to the amplification of the returning auditory feedback (Black, 1951). This greater than normal vocal intensity is not characteristic of pathological stuttering and is theoretically puzzling in other ways (see the following discussion of feedback control theory).

Generality of the Phenomenon

The repetition errors that occur under DAAF are not limited to speech. For example, expert generators of Morse code also make repetition errors, producing n + 1 instead of n *dits* or *dahs* when they hear the click of their key amplified and delayed by about 0.2 s (Smith, 1962). DAAF also causes similar errors in whistling, singing, rhythmic hand clapping, and the playing of musical instruments (Kalmus, Denes, & Fry, 1955), indicating that many perceptual–motor systems are susceptible to repetition errors and that repetition errors may represent a general effect of DAAF.

Factors Related to the Critical Delay

The most interesting fact about DAAF has proved the most difficult to explain: the existence of a critical delay that produces maximal interference. Not all delays are equally effective in disrupting fluency. For adults, disruption increases as a function of feedback delay up to about 0.2 s and then decreases with longer delays but never disappears completely, even with delays as long as 0.8 s (Figure 10.1, D. G. MacKay, 1968). In this section I examine three factors that have been shown to influence or covary with the delay that produces maximal disruption of speech. I then discuss how these factors differ from those influencing overall disruption across all delays.

Age and the Critical Delay

D. G. MacKay (1968; see also Ratner, Gawronski, & Rice, 1964) showed that the delay that produces maximal disruption of speech varies as a function of age: 0.2 s for adults, 0.375 s for children aged 7 to 9, and approximately 0.725 s for children aged 4 to 6. Figure 10.1 illustrates the shape of the function for repetition errors, but similar functions are obtained for other error types (Fairbanks & Guttman, 1958) and for other measures of rate such as "correct syllable duration," where the effect of errors has been factored out (D. G. MacKay, 1968). Although the precise location of the peak interference delay for 4- to 6-year-old children is indeterminate in Figure 10.1, other data discussed in D. G. MacKay (1968) corroborate a hypothesized peak of about 0.70 to 0.75 s. An effect of age on overall degree of disruption can also be seen in Figure 10.1, but I will leave degree-of-disruption factors for later in the chapter.

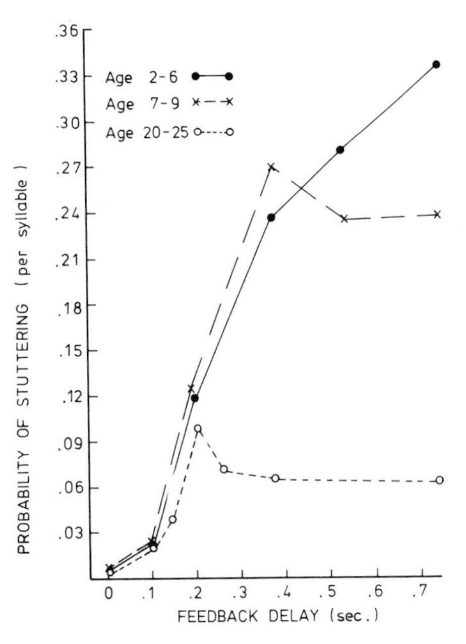

FIGURE 10.1. The probability of stuttering (per syllable) under delayed and amplified auditory feedback (DAAF) as a function of subject age and feedback delay. (Adapted from "Metamorphosis of a Critical Interval: Age-Linked Changes in the Delay in Auditory Feedback That Produces Maximal Disruption of Speech" by D. G. MacKay, 1968, *Journal of the Acoustical Society of America, 19*, pp. 816–818. Copyright 1968 by the Acoustical Society of America. Reprinted by permission.)

Potential Rate and the Critical Delay

Potential rate, operationally defined as a subject's maximum speech rate under ideal speaking conditions, is strongly correlated with the peak interference delay. D. G. MacKay (1968) first determined the potential rate for 32 subjects producing sentences with synchronous (undelayed) feedback, then had these subjects speak under DAAF, and determined the delay that produced maximal interference. Results for these two tasks were positively correlated (over $+0.5$, $p <$ .05). That is, the slower a subject's potential rate (maximum speech rate with undelayed feedback), the longer the DAAF delay required to produce maximal interference in that subject. As D. G. MacKay (1968, pp. 818–819) points out, "Clearly, some factor or set of factors limiting a subject's maximum rate of speech must determine the (temporal) locus of delayed auditory feedback interference." I argue in the following discussion that the recovery cycle of mental nodes may be one of these factors.

Syllable Repetition and the Critical Delay

Syllable repetition seems to eliminate the peak interference effect. Data published for the first time in the following section indicate that manipulating feedback delay has relatively little effect on errors or speech rate under DAAF when subjects simply repeat a single syllable as opposed to producing a normal sentence.

D. G. MacKay and Burke (1972)

D. G. MacKay and Burke (1972) conducted a series of experiments systematically comparing the effects of DAAF delay on production of repeated versus nonrepeated syllables. Subjects (N = 40) either read a 20-syllable sentence or repeated a single syllable (*puh*, *buh*, *guh*, *nuh*, or *suh*) 20 times. In both conditions, subjects spoke at maximum rate until signaled to stop, about 5.0 s after speech onset. Sound pressure level was amplified to approximately 95 dB over earphones the subjects were wearing, and feedback was delayed by either 0, 50, 100, 150, 180, 200, 250, or 300 ms on any given trial. Order of presentation of the 8 delay conditions was randomized across both subjects and materials.

The results appear in Figure 10.2. As can be seen there, repetition errors and mean syllable duration varied in the characteristic way as a function of feedback delay during sentence production. Both measures of interference increased to a maximum with a delay of about 200 ms and decreased thereafter.

Syllable repetition gave a strikingly different pattern of results, however. No errors occurred for any delay during syllable repetition, and the time to repeat a syllable remained essentially constant over the 8 delay conditions (Figure 10.2). Figure 10.2 seems to suggest the possibility of a double peak in the syllable repetition function, one peak at 50 ms and another peak at 200 ms. However, we were unable to verify existence of these peaks statistically or to replicate them in subsequent experiments.

In either of these subsequent experiments, subjects (N = 20) repeated a digit (either *two*, *four*, or *eight*) while hearing their auditory feedback amplified and

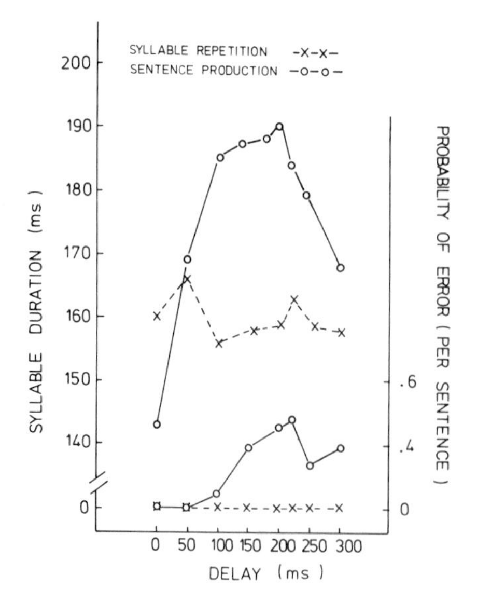

FIGURE 10.2. Syllable duration (left ordinate) and probability of error (right ordinate) for the production of sentences versus the repetition of syllables (D. G. MacKay & Burke, 1972).

delayed, and again, syllable duration remained relatively constant as a function of feedback delay. However, when the same subjects counted as rapidly as possible from 1 to 10, the 200-ms peak characteristic of connected sentences reappeared. This finding indicates that repetition per se was responsible for the flat feedback delay function in Figure 10.2 and not the nature of the syllables in the syllable repetition condition.

Another interesting difference between producing repeated versus connected syllables is the slower rate for syllable repetition in the 0-delay condition (Figure 10.2). This difference was especially marked toward the end of the 5-s production period. Syllable repetition became progressively slower over the 5 s and over the course of the experiment, as if a fatigue or satiation process (Chapter 1) were reducing the maximal rate of speech.

Factors Unrelated to the Critical Delay

Two factors *do not* influence the critical delay. These are the actual rate of speech and practice, and explaining why these factors are unrelated to the critical delay also presents a challenge for theories of DAAF.

Actual Rate of Speech

In contrast to *potential* rate, that is, a subject's maximal speech rate with synchronous feedback, *actual* (voluntarily determined) rate of speech under DAAF is unrelated to the delay that produces maximal interference. The critical delay remains unchanged when adults voluntarily speak more slowly. As can be seen in Figure 10.3 (D. G. MacKay, 1968), the frequency of repetition errors per syllable remains greatest at the 0.2-s delay for adults speaking at three different rates under two conditions of delay. This finding suggests that mechanisms determining the *actual* rate of speech must differ from those limiting the *maximal* rate (see following).

Practice (Language Familiarity)

Practice or language familiarity is another factor that seems unrelated to the peak interference delay. For example, when bilinguals produce their more and their less familiar language under DAAF, the delay that produces maximal interference with their speech remains constant (D. G. MacKay, 1970b; see also Figure 10.4).

Factors Influencing Degree of Disruption

Five factors have been shown to influence the overall degree of disruption independent of feedback delay. I have already mentioned one, level of amplification. Another factor influencing degree of disruption is actual speech rate. The law of speed–accuracy trade-off that characterizes other errors also characterizes errors under DAAF (Figures 10.3 and 10.4; see also Kodman, 1961; D. G.

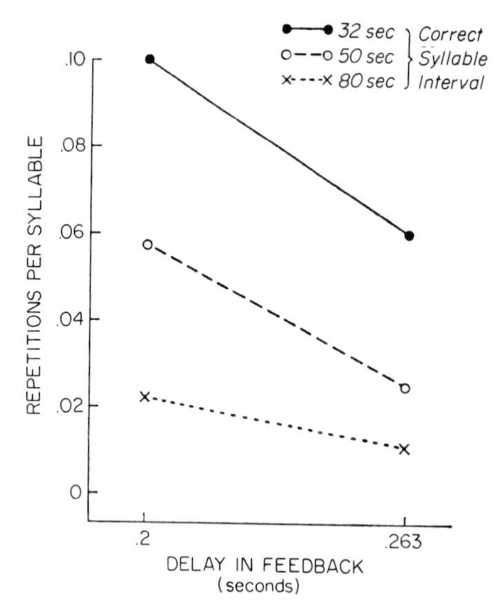

FIGURE 10.3. Repetition errors under two delay conditions for three rates of speech. (Adapted from "Metamorphosis of a Critical Interval: Age-Linked Changes in the Delay in Auditory Feedback That Produces Maximal Disruption of Speech" by D. G. MacKay, 1968, *Journal of the Acoustical Society of America, 19*, pp. 816–818. Copyright 1968 by the Acoustical Society of America. Reprinted by permission.)

MacKay, 1968; 1971). This finding indicates again that mechanisms for voluntarily speaking more slowly differ from mechanisms that limit maximal or potential speech rate. D. G. MacKay (1968) showed that errors under DAAF correlated *positively* rather than *negatively* with a subject's *potential* rate.

Another factor influencing degree of DAAF disruption is familiarity or practice with the materials being produced. For example, bilinguals stutter more when producing their less familiar language under DAAF (Figure 10.4), and

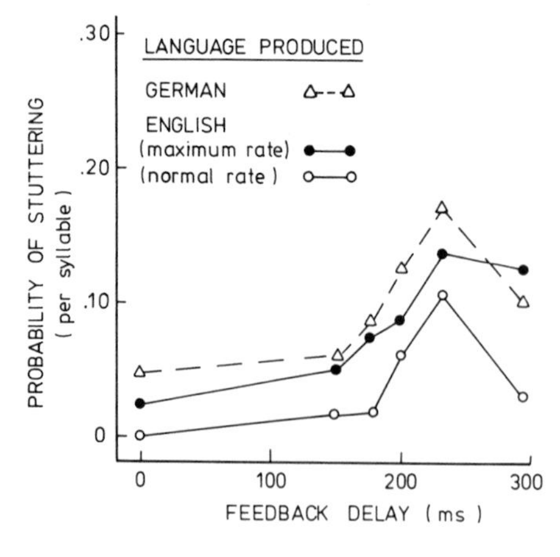

FIGURE 10.4. The probability of stuttering (per syllable) as a function of feedback delay for bilingual Americans speaking English (solid circles) and German (triangles) at their maximum rate and English at their normal rate (open circles). (Adapted from "How Does Language Familiarity Influence Stuttering Under Delayed Auditory Feedback?" by D. G. MacKay, 1970, *Perceptual and Motor Skills, 30*, p. 663. Copyright 1970 by Perceptual and Motor Skills. Reprinted by permission.)

practice in producing a sentence with synchronous feedback reduces the probability of stuttering when subjects subsequently produce the practiced sentence under DAAF (D. G. MacKay, 1970b; D. G. MacKay & Bowman, 1969). Overall disruption also diminishes with age, because practice, familiarity, or experience in producing speech increases as children grow older (Figure 10.1; D. G. MacKay, 1968).

Finally, altering either the normal muscle movements or the returning auditory feedback reduces overall interference under DAAF. D. G. MacKay (1969d) had subjects produce an "accent" by actively contracting velar muscles so as to nasalize all of their speech sounds. These unusual muscle contractions produced unaccustomed acoustic feedback and significantly reduced DAAF interference (with controls for speech rate and intensity level of the returning feedback). In a second experiment, D. G. MacKay (1969d) had subjects "passively" alter the nasal quality of their acoustic feedback by holding their nose while speaking, and this procedure also diminished the degree of DAAF interference. Hull (1952) and Roehrig (1965) used band pass filters to "passively" distort returning auditory feedback and likewise reported that DAAF disruption diminished with degree of distortion. The only experiment *not* reporting diminished DAAF disruption with feedback distortion is Howell and Archer (1984). However, their speech production task involved vowel repetition, and effects of DAAF are known to differ for repeated versus connected syllables (see preceding discussion; and Chase, 1958).

Feedback Control Theory and Feedback Disruption

I turn now to theoretical explanations of the effects of DAAF. Feedback control theories provided the initial framework and impetus for studies of DAAF but proved incapable of explaining either the overall degree of interference or the peak interference delay. This, along with some other problems discussed later, contributed to the current unpopularity of these theories in cognitive psychology (but see Holland et al., 1986; Rumelhart, Smolensky, McClelland, & Hinton, 1986, for a feedback control hypothesis involving internal feedback arising from thought or internally generated actions).

The main assumption of traditional feedback control theories is that external feedback from an ongoing action plays a direct role in controlling subsequent action, and this feedback control assumption has proven useful in describing innate regulatory behaviors such as the pupillary reflex. (See e.g., Oatley, 1978. Onset of a bright light causes reflex pupillary contraction, diminishing pupil diameter, and thereby reducing the amount of light falling on the retina, a feedback effect that in turn causes diminished contraction, until some goal or control point is reached.)

Feedback control theories have proven less useful in describing learned and highly skilled behaviors such as speech. Several different feedback control theories have been advanced for speech production (D. G. MacKay, 1969d), and all predict that distorting or eliminating feedback should cause interference. Of course, masking or distorting auditory feedback has no such effect, and feedback control theories usually assume that speakers switch to bone-conducted feedback for controlling speech under these unusual circumstances. However, speech production also remains unimpaired for many months after an injury that causes complete and total deafness (Siegel & Pick, 1974), a finding that makes it difficult to imagine that articulation requires auditory control of any kind. Articulatory adjustments for producing speech sounds have a time span of milliseconds not months.

All versions of feedback control theory also have difficulties with both the detailed and the more general effects of DAAF. For example, why is there a delay that produces maximal disruption of speech? Under feedback control theory, disruption should either remain constant or increase monotonically as a function of delay, rather than *decreasing* after some critical delay (D. G. MacKay, 1969d). Why is it necessary to *amplify* the returning feedback in order to bring about articulatory errors? Why do subjects speak *louder* when their amplified auditory feedback is delayed? After all, *undelayed* amplified feedback causes people to speak *softer* (as expected under feedback control theory; see Siegel & Pick, 1974). Why does distortion of returning auditory feedback *reduce* DAAF interference? Under feedback control theory, distortion should *increase* the difficulty of feedback control. Why does practice or familiarity with a sentence reduce DAAF interference? Under the feedback control theory of Adams (1976), practice strengthens an internal trace of the expected feedback, and successive movements are driven by the discrepancy between the ongoing feedback and the expected feedback or feedback trace. This means that practice should *increase* rather than decrease the probability of errors for sentences produced under DAAF. These and other discrepancies suggest that articulation is not under direct feedback control and that a new explanation for disruptive effects of feedback is needed (see also Siegel & Pick, 1974). The remainder of this chapter develops such an explanation.

Feedback Disruption and the Recovery Cycle

In order to underscore the intimacy of the relation between self-inhibition and the processing of feedback in the node structure theory, I will first review the potentially disruptive effects of bottom-up internal feedback, noted briefly in Chapter 8. My thesis is that preventing these potentially disruptive effects of external and internal feedback is one of the main functions of self-inhibition. However, DAAF happens to bypass this and other defense mechanisms that have evolved to prevent these disruptive effects.

Potentially Disruptive Effects of Internal Feedback

In the case of internal feedback, the problem is this. Because mental nodes receive both bottom-up and top-down connections, internal feedback can potentially cause reverberatory activation of mental nodes at every level in the system. During production, a superordinate node becomes activated and primes its subordinate nodes via top-down connections. However, the subordinate nodes become activated soon thereafter, priming their superordinate node via the bottom-up connections required for perception using exactly the same nodes. Self-inhibition is therefore needed at every level to ensure that bottom-up priming from subordinate nodes does not lead to the reactivation of just-activated nodes.

Disruptive Effects of External Feedback

Self-inhibition is also required to prevent similar effects resulting from external feedback. Like internal feedback, the external feedback that arises automatically from sensory analysis of auditory or other perceptual consequences of an action introduces bottom-up priming, which could potentially cause repeated reactivation of the lowest level mental nodes that gave rise to the feedback in the first place. To prevent such reactivations, self-inhibition must continue for a relatively long time, that is, at least as long as it takes for the lower level muscle movement nodes to become activated—for the, say, airborne feedback to reach the ears—and for the sensory analysis nodes to process the feedback and deliver priming to the mental nodes that originated the output.

By way of illustration, consider node *Z*, a low-level mental node that becomes activated and primes the muscle movement nodes that eventually give rise to auditory output. When this auditory output arrives at the ears, sensory analysis nodes automatically process this feedback and eventually prime *Z*, the node responsible for generating the output in the first place. This feedback-induced priming could result in the reactivation of *Z*, except that it arrives during *Z's* period of self-inhibition. As a consequence, *Z* cannot accumulate enough priming to become most primed in its domain and cannot become reactivated when the triggering mechanism is applied to its domain during ongoing production of the remainder of the word or sentence.

Other Defenses Against Feedback-Induced Disruption

Self-inhibition is not the only defense mechanism that has evolved to prevent disruptive effects of self-produced external feedback. Stapedial attenuation provides a similar sort of defense. The stapedius muscle in our middle ear contracts just before we begin to speak and remains contracted throughout the period of speech, thereby attenuating the amplitude of eardrum vibration in response to hearing one's own voice.

Stapedial attenuation is sometimes viewed as a feedback-induced reflex for preventing damage to the eardrum that might otherwise arise from prolonged screaming. However, preventing injury cannot be the sole purpose of stapedial attenuation (see also Simmons, 1964). For example, stapedial activity also accompanies everyday speech and whispering at levels that cannot possibly cause peripheral damage. Nor is stapedial attenuation an externally triggered reflex. Borg and Zakrisson (1975) were able to actually see, as well as electromyographically record, stapedial attenuation in otherwise normal speakers with a perforated eardrum, and they found that stapedial contraction preceded vocalization by about 75 ms (even when subjects spoke with very low acoustic intensity). Stapedial attenuation must therefore arise from a central command that precedes vocalization, rather than from a peripheral reflex triggered by vocal feedback. An interesting possibility under this "central command hypothesis" is that stapedial attenuation will precede and accompany production of internal speech in the complete absence of auditory feedback from the voice (Chapters 8 and 9).

Stapedial attenuation has other consequences under the node structure theory in addition to providing defense against disruptive effects of external feedback. For example, stapedial attenuation adds to the list of differences between perception of self-produced versus other-produced speech and represents another factor in the conspiracy of factors (together with self-inhibition and bone-conducted feedback) that contribute to the fact that one's own voice sounds differently during ongoing speech production than during a subsequent replay of the same sounds via tape recorder.

The Recovery Cycle Explanation of DAAF Interference

Under the recovery cycle hypothesis, feedback plays no direct role in controlling the form of a highly skilled behavior, and DAAF has its effects by overcoming the defenses against feedback-induced disruption of action. Under this view, the exact duration of the feedback delay is critical. DAAF must arrive at the point in time when mental nodes originally responsible for the output and feedback are especially susceptible to reactivation. Feedback arriving at the time when these just-activated mental nodes are self-inhibited can have no effect. To cause significant interference, DAAF must arrive slightly later, during the hyperexcitable phase of these just-activated nodes, when their level of priming is already greater than normal.

As noted in Chapter 8, the pattern of segment repetition in the structure of words suggests that hyperexcitability peaks at about 200 ms after a phonological segment node has been activated and returns to normal or spontaneous level by about 300 ms following activation.

The maximal influence of DAAF with a delay of about 0.2 s therefore reflects an effect of feedback-induced priming arriving with sufficient strength at the critical 0.2-s period in the recovery cycle of just-activated nodes. Amplification of the returning feedback adds further to the priming of the just-activated nodes, and when these combined sources of priming exceed the top-down priming for the appropriate nodes in the same domain, these just-activated nodes are automatically reactivated under the most-primed-wins principle, so that the output resembles stuttering (D. G. MacKay & MacDonald, 1984). In order to simplify exposition, I have, of course, ignored the time it takes for muscles to move following activation of a phonological node, the time it takes airborne auditory feedback to arrive at the ears, and the time it takes sensory analysis nodes to process the feedback and deliver bottom-up priming to the just-activated phonological node.

Under this account, bottom-up priming arising from DAAF causes repetition errors directly and causes substitution errors indirectly via summation with top-down priming being transmitted to the intended or appropriate phonological nodes. If this summated top-down and bottom-up priming happens to make a new combination of distinctive feature nodes most primed in their domains when their activating mechanisms are applied, speakers could produce a phoneme that is not part of any language familiar to them, just as B. S. Lee (1950) observed.

The fact that subjects speak louder at the most disruptive delays (Black, 1951) may reflect an attempt to augment top-down priming and thereby enable the appropriate nodes to dominate in the most-primed-wins competition with bottom-up priming from DAAF. Transfer effects of practice or language familiarity (D. G. MacKay, 1982) are explained in a similar way and follow directly from the linkage strength assumption of the node structure theory. That is, practice without DAAF reduces interference when a sentence is later produced under DAAF because linkage strength of top-down connections will increase, thereby enabling top-down priming to compete more effectively against bottom-up priming from DAAF.

Consider now the factors influencing the delay that produces maximal interference. Under the recovery cycle hypothesis, two underlying processes determine the peak interference delay. The main one concerns the temporal characteristics of rebound hyperexcitability in sequence and content nodes. This rebound from self-inhibition is automatic but individual specific, varying with individual-specific factors such as age and experience (see following discussion). Rebound from self-inhibition is unrelated to processes governing the *actual* (voluntarily specified) rate of speech, determined by the oscillation rate of timing nodes.

However, temporal characteristics of recovery from self-inhibition *do* influence a subject's *potential* rate (maximal speech rate with synchronous feedback). Potential rate is determined in part by the rate at which low-level nodes can be reactivated (for a given error criterion, see D. G. MacKay, 1982), and this reactivation rate is influenced in turn by the time characteristics of recovery from self-inhibition. This means that potential speech rate and time of peak hyperexcitability will be positively correlated, explaining D. G. MacKay's (1968) observation that the faster a subject's potential rate, the shorter the critical delay that produces maximal interference under DAAF. However, because factors influencing the time course of recovery from self-inhibition are central, automatic, and beyond voluntary control (see following), one would not expect the delay that produces maximal interference to shift when subjects *voluntarily* speak slower or prolong speech sounds, and no such shift is found (D. G. MacKay, 1968; 1970b).

The rate at which a node can be reactivated is also related to the level of priming that the node achieves, relative to all other nodes in its domain, and degree of satiation of the node directly influences this level of priming. Satiation therefore explains the slower maximal speech rate for producing repeated versus connected syllables, as well as the decrease in repetition rate with increasing numbers of repetitions (see preceding discussion of D. G. MacKay & Burke, 1972).

By reducing the overall level of priming of a node, satiation also increases the overall probability of errors, without influencing either the hyperexcitability peak, or the delay that produces maximal interference. Some other factor therefore is needed to explain the fact (discussed earlier) that repeating a syllable under DAAF proceeds without errors and at a rate that does not vary as a function of delay.

PRACTICE AND THE PEAK INTERFERENCE DELAY

The recovery cycle hypothesis predicts a positive correlation between prior practice and the delay that produces maximal interference under DAAF. Recall from Chapter 8 that practice speeds up the activation and recovery cycle of individual nodes by strengthening the connection between the parent node and its inhibitory satellite. Onset of hyperexcitability will therefore vary with a node's history of prior activation, and hyperexcitability peaks much longer than 200 ms can be expected for unpracticed phonological nodes.

Practice-induced speed-ups in the activation and recovery cycle readily account for the longer delays required to produce peak interference in 2- to 6-year-old children (Figure 10.1) but seem to fly in the face of my own data for bilinguals (Figure 10.4). D. G. MacKay (1970b) reported that peak interference delay remained constant when German–English bilinguals produced their more and their less familiar language under DAAF. However, all of the native German speakers in this experiment had *at least* 5 years of intensive prior practice with English, and some had more than 20 years of prior practice. And because phonological nodes receive so much practice so quickly (D. G. MacKay, 1982), phonological nodes for these subjects must have already reached asymptotic levels of practice for both languages (especially in view of the extensive overlap between phonological components for German and English). And because phonological nodes provide the "first line of defense" against disruptive effects of feedback, no differences in peak interference delay for the more versus less familiar language could be expected in D. G. MacKay (1970b). However, the relationship between prior practice and peak interference delay warrants further test. The node structure theory predicts peak interference delays much longer than 200 ms for English speakers learning, say, Swahili, a language with phonological units very different from English.

Feedback-Induced Stuttering and the Recovery Cycle Hypothesis

What follows is another in a long history of attempts to provide an account of feedback-induced stuttering that integrates stutterers and nonstutterers. Under the recovery cycle hypothesis, the muscle movement nodes of stutterers display

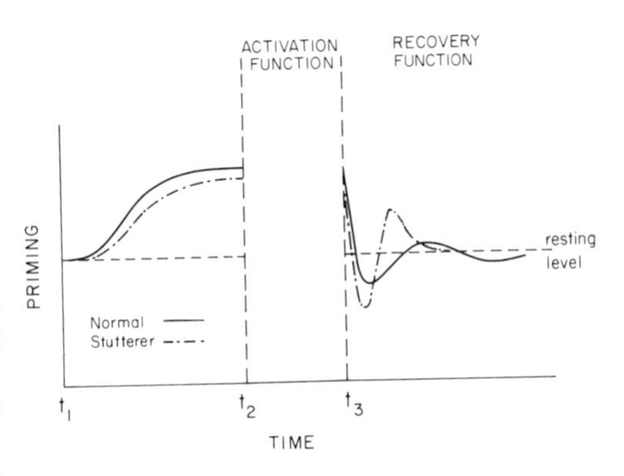

FIGURE 10.5. The priming function (relating degree of priming and time following onset of priming at t_1), activation function (following application of the triggering mechanism at t_2), and recovery function (following activation offset, t_3) for stutterers and nonstutterers. (Adapted from "Stuttering as a Sequencing and Timing Disorder" by D. G. MacKay and M. MacDonald in *Nature and Treatment of Stuttering: New Directions* (p. 273) edited by W. H. Perkins and R. Curlee, 1984, San Diego: College-Hill. Copyright 1984 by College Hill Press. Reprinted by permission.)

an abnormal recovery cycle, as illustrated in Figure 10.5. That is, rebound from self-inhibition comes earlier than normal and rises to a higher level of priming. As a result, just-activated nodes in their hyperexcitability phase may have greatest priming in their domain at the time when the next node is to be activated.

Heightened hyperexcitability alone could cause repetition errors, but amplifying the auditory feedback further increases the probability of stuttering. The delayed or attenuated stapedial contraction that has been observed in (feedback-induced) stutterers (Hall & Jerger, 1978; Horovitz, Johnson, & Pearlman, 1978) will likewise provide less of a defense against feedback-induced stuttering. Unattenuated external feedback will return with greater than normal amplitude to the just-activated mental nodes and further increase the probability of reactivation.

Mental nodes and the recovery cycle hypothesis explain why masking the returning auditory feedback reduces the probability of stuttering (see Baar & Carmel, 1970; Cherry & Sayers, 1956; and findings discussed below), and amplifying it has the opposite effect in feedback-induced stutterers (findings discussed below). Mental nodes also make sense of auditory induced fluency, the fact that appropriate auditory input can guide fluent production of a word, causing release from blocks and prolongations. Blocks are overcome when someone else utters the word on which a stutterer is blocking, because this input primes the appropriate or next-to-be-activated nodes, enabling these nodes to achieve greatest priming in their domain and become activated. Shadowing and choral rehearsal likewise prevent stuttering, because these auditory inputs augment priming for the appropriate or next-to-be-activated nodes.

One interesting prediction from the recovery cycle hypothesis is that the delay that produces maximal disruption with speech under DAAF will be related to potential rate, that is, to a stutterer's maximal articulatory rate without DAAF. The reasoning is already familiar. Potential rate is determined in part by the maximal rate at which nodes can be reactivated, which by hypothesis is determined by the speed of recovery following activation of these nodes. Thus, a stutterer whose maximal rate of speech is relatively fast can be expected to display a shorter period of self-inhibition, with faster rebound and shorter peak interference delay under DAAF than a stutterer whose maximal rate of speech is relatively slow.

Another interesting prediction is that feedback-induced stutterers will react differently from intrinsic stutterers and normals to DAAF. Specifically, the delay that produces maximal interference with speech will be shorter for feedback-induced stutterers than for intrinsic stutterers and normals under the recovery cycle hypothesis. Further research is needed to test this prediction, because systematic comparisons of the effects of different delays on speech of normals, intrinsic stutterers; and feedback-induced stutterers' have never been undertaken. Previous DAAF studies have lumped intrinsic and feedback-induced stutterers together and either have used only a single feedback delay, have failed to amplify returning feedback, have omitted the normal control group, or have neglected basic controls for speech rate, distraction effects, order of the delays, and possible practice effects over repeated readings of the same materials (D. G. MacKay & MacDonald, 1984).

A Test of the Recovery Cycle Hypothesis

An experiment by D. G. MacKay and Birnbaum, reported for the first time here, incorporated all of the necessary control procedures just noted (for details, see D. G. MacKay, 1970b). The experiment examined effects of DAAF on three groups of subjects of about the same average age: feedback-induced stutterers who read a set of sentences more fluently when hearing white noise than when not hearing white noise (see Table 10.1), intrinsic stutterers who read as fluently or less fluently when hearing white noise than when not hearing white noise (see Table 10.1), and nonstutterers who, of course, read fluently under either noise or no-noise conditions. In the main experiment, the subjects read sentences as quickly as possible with auditory feedback amplified to about 95 dB under 10

TABLE 10.1. The time (in seconds) to read sentences (10 ± 1 syllables in length) under two conditions of undelayed feedback and two conditions of amplification by two groups of stutterers (see text for explanation).

Feedback conditions by stutterer group	White noise	No white noise
Feedback-induced stutterers		
Loud feedback	2.44	3.88
Soft feedback	2.43	3.01
Intrinsic stutterers		
Loud feedback	3.92	3.58
Soft feedback	3.20	3.24

different conditions of feedback delay (0.0 s, 0.01 s, 0.025 s, 0.075 s, 0.150 s, 0.200 s, 0.250 s, 0.400 s, 0.550 s, and 0.800 s). Effects of these delays were as follows. Peak interference delay (measured via either errors or syllable duration) always exceeded 0.01 s and differed among the three groups. The delay that produced maximal disruption was 0.2 s for the nonstutterers (as usual) and was consistently longer than for the *feedback-induced* stutterers (median 0.150 s) with no overlap in the distributions. This finding strongly supports the prediction of the recovery cycle hypothesis that peak interference delay is shorter for feedback-induced stutterers than for nonstutterers.

Results for the *intrinsic* stutterers were strikingly different. For this group, the delay that produced maximal interference was *longer* than for nonstutterers, that is 0.400 s for intrinsic stutterers versus 0.200 s for nonstutterers. This difference further corroborates the distinction between intrinsic versus feedback-induced stuttering and is reminiscent of the effects of DAAF in children. The peak interference delay is 0.4 s in 7- to 9-year-old children, and longer than in adults (0.2 s) (D. G. MacKay, 1968). One therefore wonders whether a development deficit may play a role in intrinsic stuttering, such that the age-linked shift in peak interference delay (and by hypothesis the faster activation and recovery cycle of underlying nodes) has failed to occur.

Epilogue

> To help with the division of labor in a field of this scope, researchers have adopted what might be called the dichotomization strategy. Commonsense dichotomies having intuitive or practical appeal rather than theoretical significance are used to segregate the field, and create subfields with more manageable research literatures.
>
> (D. G. MacKay, 1982, p. 485)

> During the late 1800s and early 1900s, leading psychologists such as Wundt, Liepmann, Sherington, and Thorndike actively pursued the topic of action and motor performance. With the rise of cognitive psychology, the focus shifted. . . . Perhaps motor performance was considered too simple to be interesting, or perhaps it did not have the intellectual aroma of topics like concept formation or memory of visual perception. The little research in motor performance was usually relegated to departments of physical education or kinesiology, where it had little contact with mainstream psychology.
>
> (Gentner, 1985, p. 183)

A recent overview of the accomplishments and shortcomings of experimental psychology over the last hundred years (D. G. MacKay, in preparation) argues that the failure to develop general and plausible theories is our main shortcoming, with noncumulative development of facts being one of the main side effects of this shortcoming. Without becoming integrated into a coherent theory, empirical phenomena are frequently set aside and forgotten. The present book can be viewed as an attempt to rescue some of these forgotten phenomena, among them the various manifestations of self-inhibition, including stuttering and the interference with speech production and other skilled behaviors that occur when auditory feedback is amplified and delayed by about 0.2 s.

To some readers, stuttering may seem a somewhat peripheral point on which to end a book entitled *The Organization of Perception and Action*. However, stuttering research illustrates in microcosm some of the more general problems endemic to the field at large, and in this Epilogue, I examine these problems and how to solve them. I then summarize what I feel are some of the major contributions and weaknesses of the book.

Stuttering and Theoretical Psychology

Stuttering is important for theories of normal behavior, because a complete and adequate theory must be capable of explaining all of the ways that an output system will break down, and as a breakdown with a perceptual cause, feedback-induced stuttering is especially important for theories of the relation between perception and action.

The history of stuttering research also carries important implications for the field at large. Current problems in psychology, such as fragmentation of the field, misunderstandings as to the nature of theories, atheoretical approaches with possibly self-perpetuating effects, and ahistoricity or noncumulative development of knowledge (D. G. MacKay, in preparation), are all writ large in the 120-year history of stuttering research (which begins with Wyneken, 1868). I argue here that these problems with stuttering research can be traced to a general "divide-and-conquer" strategy that has been adopted by the field at large.

The Divide-and-Conquer Approach in Stuttering Research

Over the past several decades, psychology has been following what might be called the divide-and-conquer approach to theory construction. Under this approach, a subdomain of the field is segregated on practical or intuitive grounds in order to develop one or more unique empirical approaches and experimental paradigms for generating a (hopefully) coherent body of facts, insights, and "miniature" theories within the subdomain. Given this coherent body of facts and insights, the goal is then to reintegrate the subdomain with the field at large, to the benefit of all concerned.

Stuttering research clearly illustrates the motivation underlying the divide-and-conquer approach. On the one hand, stuttering seems different from any other behavior that psychologists are interested in (see the discussion of differences between stuttering and other speech errors in Chapter 8), and on the other hand, mainstream psychology seems to offer little to researchers who want to understand stuttering and perhaps also to help provide relief for stutterers. As Van Riper (1982) points out after a lifetime of work in this area, psychological theories of speech production with applications to stuttering have been "slow in coming."

Stuttering research also illustrates a successful first step in the divide-and-conquer approach: *division*. Stuttering has been studied as a kingdom apart, by and large independently of what psychology has discovered about normal speech production, and many researchers now consider stuttering a field unto itself, with its own special phenomena, methodology, and theories (D. G. MacKay & MacDonald, 1984).

What about the *conquer* aspect of the divide-and-conquer approach? Has a coherent body of facts and insights about stuttering emerged and become part of the field at large? Absolutely not. Despite the practical and theoretical significance of stuttering, and despite the thousands of empirical studies that have examined stuttering under the divide-and-conquer approach, mainstream psychology and stuttering research have continued to go their separate ways (Garber & Siegel, 1982). As in psychology at large, the divide-and-conquer strategy has also retarded the development of theory. Most theoretical ideas about stuttering have been either descriptive in nature or so vague as to be unhelpful (Bloodstein, 1969). For example, "theories" that call stuttering a perseverative response, a symbolic sucking activity, or a miniature convulsion at best only loosely describe rather than explain it. Similarly, attributing stuttering to conflict, anxiety, stress, or delayed myelinization of cortical neurons is theoretically unhelpful unless a detailed causal explanation can be provided for at least one specific, real-time example of stuttering behavior.

More generally, the problem is that theories developed under the divide-and-conquer approach are at best a stab in the dark and at worst simply not possible. In particular, it is not possible to construct a genuine theory that attempts to explain stuttering independently of the mechanisms for producing error-free speech. Although stuttering is important for mainstream psychology, it is equally true that an understanding of the processes underlying normal behavior is necessary for understanding its disruption in a complex disorder such as stuttering. A genuine theory must simultaneously explain how stutterers speak fluently, when they do, and how fluency breaks down, when it does. A miniature theory that attempts to explain stuttering per se is analogous to a hypothetical theory in physics that applies only to backfires emitted from the exhaust system of a car. One cannot explain backfires independently of the principles of internal combustion underlying the normal functioning of an automobile engine. Similarly, one cannot explain stuttering independently of the mechanisms underlying the production of error-free speech (D. G. MacKay, 1969b; 1970a). Nor is this problem with miniature theories confined to stuttering research. As Lachman, Lachman, and Butterfield (1979) point out, the less than satisfactory nature of miniature theories developed under the divide-and-conquer approach seems to represent a general problem in psychology, which it is important, if possible, to correct (see also D. G. MacKay, in preparation).

In summary, the divide-and-conquer approach to stuttering has *in principle* frustrated the development of theory. No coherent account of stuttering either has emerged or can emerge under the divide-and-conquer strategy, and even though large numbers of findings have been amassed, many have been forgotten. Bridging the longstanding gap created by the divide-and-conquer approach between mainstream psychology and stuttering research is not just desirable but necessary for a solution to the riddle of stuttering. The "deplorable" state of stuttering research (Preus, 1981) can be expected to continue as long as stuttering research remains separate from theoretical psychology at large.

The lesson here, as elsewhere in psychology (D. G. MacKay, in preparation), is that general theories are in order and an approach that limits a field to the accumulation of empirical findings should be abandoned. What is needed in addition to further experiments is a genuinely theoretical psychology for integrating available observations and pointing the way to empircal regularities that are significant and new (see also D. G. MacKay, in preparation).

The Node Structure Theory in Overview

What new and significant regularities does the node structure theory point to? Perhaps the main ones are summarized by the principle of higher level activation. One of the distinctive characteristics of the node structure theory is its differentiation between content nodes (the units representing the basic components of action and perception), sequence nodes (the units that activate and sequence the basic components), and timing nodes (the units that determine when and how rapidly the basic components become activated). The general processing characteristics of these nodes (e.g., priming and activation); their basic structural properties (e.g., hierarchical organization with respect to activation and heterarchical organization with respect to priming), and their long-term memory characteristics (e.g., linkage strength) are not particularly new (Chapter 1). However, the precise nature of the interactions between these processing characteristics in the theory provide the basis for new theoretical generalizations such as the principle of higher level activation, which summarizes a large number of empirical regularities, from perceptual invariance to perception of the distal stimulus (Chapter 4).

Another not-so-new set of empirical regularities that acquires significance under the node structure theory concerns the asymmetries discussed in Chapter 6 between the potential rate of perception and action (the maximal rate asymmetry); between effects of practice on perception versus production (the listening practice asymmetry); between the ease of learning to recognize versus produce words (the word production asymmetry); and asymmetries between errors in perception versus production.

A final, but again not entirely new set of empirical regularities given significance under the node structure theory concerns the many parallelisms between perception and production (in both units and empirical effects, Chapter 2) and the many interactions (both facilitative and disruptive) between the processes for perception versus production (e.g., in the organization of timing, Chapter 5).

More specific contributions of the theory include new hypotheses concerning the perceptual processes underlying categorical perception and its exceptions, visual dominance and its exceptions, perceptual errors, and the processing of ambiguous inputs. The theory also extends existing perceptual hypotheses about the units of perception and the explanation of right-to-left effects (as in phonological fusions) and left-to-right effects (such as the greater ease of detecting "mispronunciations" at the beginnings than at the ends of words). Other new hypotheses concern constant relative timing, effects of practice on the timing and sequencing of behavior, and regularities in the nature of errors (e.g., speed–accuracy trade-off, the sequential class regularity and its exceptions, and the level-within-a-system effect, which is the fact that higher level units within a system are more prone to error than lower level units within the same system). The theory also develops new hypotheses about the functions of stapedial attenuation, the role of internal and external feedback in detecting and correcting errors (both overt and internally generated), and differences between the perception of self-produced versus other-produced speech (e.g., the size-of-unit effect in the detection of errors and the missing feedback effect in verbal transformation experiments). Finally, the theory provides a detailed and testable account of stuttering and the disruptive effects of delayed and amplified auditory feedback.

Comparisons With Other Current Theories

How does the node structure theory relate to other current theories? As noted in the Introduction, the node structure theory is much broader in scope than other theories and treats perception and action not as separate but as closely interrelated processes involving the same higher level mechanisms and structures (the so-called perception-and-action approach referred to in the Preface). However, like other current theories, the node structure theory participates in the current trend toward a focus on dynamic, on-line, or real-time process issues, in addition to static or structural issues. Like other recent theories (e.g., Grossberg, 1982), the node structure theory also postulates underlying mechanisms and processes that are within the feasibility of simple neuronal or neuronlike circuits. The theory is physiologically plausible or has the potential for mapping psychological constructs onto neuroanatomical constructs (see also D. G. MacKay, 1985; McClelland et al., 1986), an exciting prospect, because, as Ojemann (1983) points out, some such mapping seems essential for cracking the code of the brain. Indeed, the node structure theory may eventually rise or fall depending on physiological evidence for constructs resembling inhibitory satellites, sequence nodes, and domains and systems of content nodes.

Earlier chapters have compared the node structure theory with other recent theories of perception and/or action on a number of more specific dimensions, among them, "mass action" versus central, localized, or local processes and representations, the representation of phonological rules and phonological units in general, representations of ambiguity and "shades of meaning," the representation of conscious versus unconscious processes, the processing of perceptual feedback and its use in error correction and motor control, sequencing principles and parallel versus serial processing in general, sequential rules versus condition–action rules, timing principles or the lack thereof, ways in which action influences perception, top-down effects in perception and bottom-up effects in production, categorical versus noncategorical perception. However, I have not attempted to compare the node structure theory with any other theory overall and in detail. The reason is not just that such an endeavor would double the length of an already lengthy book (which it would if well done). Detailed overall comparisons will only be really useful when the field has developed two or more well-established theories of equivalent scope and precision. Theoretical psychology is nowhere near that point. Like other current theories, the node structure theory is not well established, has obvious strengths and weaknesses, and is in a state of evolution. Much more will be gained by direct attempts to overcome its weaknesses than by soon-to-be-out-of-date comparisons with other theories (including itself: for example, note the many differences between the present theory and D. G. MacKay, 1982, or between McClelland & Elman, 1986, and Elman & McClelland, 1984).

Limitations of the Node Structure Theory

I turn now to the areas where the node structure theory requires further work: formalization and scope.

Formalization of the Theory

D. G. MacKay (in preparation) argues that there is a natural sequence to theory construction that can be seen in the advanced sciences such as physics. Mathematical formalisms normally follow rather than precede the development of intuitively comprehensible theory, and breadth of theory initially takes precedence over precise but paradigm-specific simulations. I have so far simulated only limited aspects of the node structure theory (D. G. MacKay, 1982), but other aspects of the theory seem ready for real-time simulations of the sort described by McClelland and Elman (1986). Computer simulations of predictions derived from the high-interactivity situations discussed in Chapters 3 to 7 are very much needed, but as noted in the Preface, the relative noninteractiveness (locality or self-containedness) of self-inhibition (Chapters 8 to 10) will make the theory's predictions easier to simulate and check against incoming data in that area.

Conceptual Extensions of the Theory

Although the node structure theory is much broader in scope than other theories, it is nevertheless too narrow to apply to much of what goes on in either everyday cognition or laboratory experiments. Unless detailed mechanisms for attention, awareness, creativity, and learning (the formation of connections) are incorporated into the node structure theory, its applicability (simulated or not) to real data and to real behavior will be severely limited. Indeed, one (anonymous) reviewer suggested that the node structure theory will remain pretheoretical until the learnability of its node structures is determined (personal communication, Nov. 1985; but see D. G. MacKay, 1987).

The node structure theory must also represent visually guided behaviors in greater detail to qualify as a *general* theory of perception and action. Although I do discuss behaviors such as piano playing, typing, and the generation of Morse code, the theory is most detailed when it comes to the perception–production of speech. I therefore agree with McClelland and Elman (1986, p. 7), in reference to their own speech-centered theory, that "we would hope that the ways in which it deals with the challenges posed by the speech signal are applicable to other domains." But to argue that the principles of speech perception–production apply more generally to other perception–production systems is not enough; details of the similarities and differences need to be worked out.

The node structure theory is also limited with regard to speech. For example, its upper bound has been the sentence and its lower bound, the phonological feature. It is silent on the complex processes that intervene between sensory receptors (such as the basilar membrane) and the phonological features for speech perception and between phonological features and the muscles for producing speech sounds. It is even silent on the (in some ways more important) processes intervening between linguistic and nonlinguistic representations of the world.

Another limitation of the theory concerns its essentially psychological focus. Its interface with neurophysiology remains largely unexplored. I currently have no solid answer to questions such as "How are timing, sequence and content nodes instantiated in the brain?" The computational adequacy of the theory from a linguistic point of view also remains to be explored. I simply do not know what additional theoretical mechanisms will be required for producing and comprehending all possible naturally spoken sentences in English or any other language.

In view of these weaknesses and the evolving state of the node structure theory (in some ways another weakness, D. G. MacKay, in preparation), perhaps the most enduring contribution of the present book will be its analysis of constraints that future theories of perception and action must explain. These constraints form a relatively cohesive set, including theoretical constraints imposed by the units involved in perception and action, the nature of sequencing and sequencing errors in perception and action, processing asymmetries between perception and action, timing interactions between perception and action, the nature of error detection and correction, and finally, disruptive effects of feedback. New constraints will undoubtedly be discovered in the future, but the set outlined here will remain as a challenge to theoretical psychology for many years to come.

References

Adams, J.A. (1976). Issues for a closed loop theory of motor learning. In G.E. Stelmach (Ed.), *Motor Control: Issues and trends* (pp. 87–107). New York: Academic Press.

Allen, G.D. (1972). The location of rhythmic beats in English: 1. *Language and Speech*, An experimental study, *15*, 72–100.

Allport, D.A. (in press). Selection for action. In H. Heuer & A. Sanders (Eds.), *Perspectives in perception and action*. Hillsdale, NJ: Lawrence Erlbaum.

Anderson, J.R. (1976). *Language, memory and thought*. Hillsdale, NJ: Erlbaum.

Anderson, J.R. (1980). *Cognitive psychology and its implications*. San Francisco: Freeman.

Anderson, J.R. (1983). *The architecture of cognition*. Cambridge, MA: Harvard University Press.

Baars, B.J., Motley, M.T., MacKay, D.G. (1975). Output editing for lexical status in artificially elicited slips of the tongue. *Journal of Verbal Learning and Verbal Behavior, 14*, 382–391.

Barr, D.F., & Carmel, N.R. (1970). Stuttering inhibition with high frequency narrow band masking noise. *Journal of Auditory Research*, *10*, 59–61.

Biederman, I. (1972). Perceiving real world scenes. *Science*, *177*, 77–79.

Bierwisch, M. (1985). *Language production and comprehension: Two sides of the same coin? (Reprinted from Kursbuch*, 1966, *5*, 77–152. W. Deutsch (Trans.) (Report No. 72). Research Group on Perception and Action. Center for Interdisciplinary Research (ZIF). West Germany: University of Bielefeld.

Bizzi, E., Dev, P., Morasso, P., & Polit, A. (1978). Effect of load disturbances during centrally initiated movements. *Journal of Neurophysiology, 39*, 542–556.

Black, J.W. (1951). The effects of delayed side-tone upon vocal rate and intensity. *Journal of Speech and Hearing Disorders*, *16*, 56–60.

Bloodstein, O. (1969). *Handbook of stuttering*. Chicago: National Easter Seal Society for Crippled Children and Adults.

Blumstein, S. (1973). *A phonological investigation of aphasic speech*. The Hague: Mouton.

Bock, J.K., & Warren, R.K. (1985). Conceptual accessibility and syntactic structure in sentence formulation. *Cognition*, *21*, 47–67.

Bond, Z.S., & Garnes, S. (1980). Misperceptions of fluent speech. In R.A. Cole (Ed.), *Perception and production of fluent speech* (pp. 115–132). Hillsdale, NJ: Erlbaum.

Bond, A.S., & Small, L.H. (1984). Detecting and correcting mispronunciations; A note on methodology. *Journal of Phonetics*, *12*, 279–283.

Borg, E., & Zakrisson, J. (1975). The activity of the stapedius muscle in man during vocalization. *Acta Otolaryngologica, 79*, 325–333.

Borger, R. (1963). The refractory period and serial choice-reactions. *Quarterly Journal of Experimental Psychology, 15*, 1–12.

Brand, J. (1971). Classification without identification in visual search. *Quarterly Journal of Experimental Psychology, 23*, 178–186.

Bregman, A.S., & Campbell, J. (1971). Primary auditory stream segregation and perception of order in rapid sequences of tones. *Quarterly Journal of Experimental Psychology, 25*, 22–40.

Brenner, N.C., Perkins, W.H., & Soderberg, G.A. (1972). The effect of rehearsal on frequency of stuttering. *Journal of Speech and Hearing Research, 15*, 483–486.

Bridgeman, B. (1986). *Biology of behavior and mind*. New York: Wiley.

Broadbent, D.E. (1971). *Decision and stress*. London: Academic Press.

Browman, C.P. (1980). Perceptual processing: Evidence from slips of the ear. In V.A. Fromkin (Ed.), *Errors in linguistic performance: Slips of the tongue, ear, pen and hand* (pp. 213–230). New York: Academic.

Bruner, J. (1957). On perceptual readiness. *Psychological Review, 64*, 123–152.

Carroll, D.W. (1986). *Psychology of Language*. Belmont, CA: Brooks/Cole.

Chang, H.L. (1959). The evoked potentials. In J. Field (Ed.), *Handbook of physiology* (Vol. 1) (p. 299). Washington, D.C.: American Physiological Society.

Chase, R.A. (1958). Effect of delayed auditory feedback on repetition of speech sounds. *Journal of Speech and Hearing Disorders, 23, 5*, 583–590.

Cherry, C., & Sayers, B.McA. (1956). Experiments on the total inhibition of stammering by external control, and some clinical results. *Journal of Psychosomatic Research, 1*, 233–246.

Chomsky, N. (1957). *Syntactic structures*. The Hague: Mouton.

Chomsky, N., & Halle, M. (1968). *The sound pattern of English*. New York: Harper & Row.

Clark, E.V., & Hecht, B.F. (1983). Comprehension, production and language acquisition. *Annual Review of Psychology, 34*, 325–349.

Clark, H.H., & Clark, E.V. (1977). *Psychology and language*. New York: Harcourt Brace Jovanovich.

Cohen, A. (1967). Errors of speech and their implication for understanding the strategy of language users. *Zeitschrift fuer Phonetik, 21, 22*, 177–181.

Cole, R.A., & Scott, B. (1974). Toward a theory of speech perception. *Psychological Review, 81*, 348–374.

Collins, A.M., & Quillian, M.R. (1969). Retrieval time from semantic memory. *Journal of Verbal Learning and Verbal Behavior, 8*, 240–247.

Colthart, M., & Funnell, E. (1987). Reading and writing: One lexicon or two? In A. Allport, D.G. MacKay, W. Prinz, & E. Scheerer (Eds.), *Language perception and production: Relationships between listening, speaking, reading, and writing* (pp. 313–340). London: Academic.

Cooper, M.H., & Allen, G.D. (1977). Timing control accuracy in normal speakers and stutterers. *Journal of Speech and Hearing Research, 20*, 55–71.

Cooper, W.E. (1979). *Speech perception and production: Studies in selective adaptation*. Norwood, NJ: Ablex.

Cooper, W.E., Billings, D., & Cole, R.A. (1976). Articulatory effects on speech perception: A second report. *Journal of Phonetics, 4*, 219–232.

Cooper, W.E., Blumstein, S.E., & Nigro, G. (1975). Articulatory effect on speech perception: A preliminary report. *Journal of Phonetics, 3*, 87–98.

Cooper, W.E., & Nager, R.M. (1975). Perceptuo-motor adaptation to speech: An analysis of bisyllabic utterances and a neural model. *Journal of the Acoustic Society of America*, *58* (1), 256–265.

Cooper, W.E., & Paccia-Cooper, J. (1980). *Syntax and speech*. London: Harvard.

Cooper, W.E., & Zurif, E.B. (1983). Aphasia: Information-processing in language production and reception. In B. Butterworth (Ed.), *Language production* (Vol. 2). London: Academic.

Cross, D.E., & Luper, H.L. (1979). Voice reaction time of stuttering and nonstuttering children. *Journal of Fluency Disorders*, *4*, 661–668.

Cutler, A. (1982). The reliability of speech error data. In A. Cutler (Ed.), *Slips of the tongue and language production* (pp. 7–28). Amsterdam: Mouton.

Cutler, A. (1985, July). *Perceptually determined production constraints?* Paper presented at the Conference on common processes in speaking, listening, reading, and writing, at the Center for Interdisciplinary Research (ZiF), University of Bielefeld, West Germany.

Cutler, A., Mehler, J., Norris, D., & Sequi, J. (1983). A language-specific comprehension strategy. *Nature*, *304*, 159–160.

Cutting, J.E., & Day, R.S. (1975). The perception of stop-liquid clusters in phonological fusion. *Journal of Phonetics*, *3*, 99–113.

Cutting, J.E., & Rosner, B.S. (1974). Categories and boundaries in speech and music. *Perception & Psychophysics*, *16*, 564–570.

Darian-Smith, I., Sugitani, M., & Heywood, J. (1982). Touching textured surfaces: Cells in somatosensory cortex respond both to finger movement and to surface features. *Science*, *218*, 906–909.

Day, R.S. (1968). *Fusion in dichotic listening*. Unpublished doctoral dissertation, Stanford University.

Dell, G.S. (1980). *Phonological and lexical encoding in speech production: An analysis of naturally occurring and experimentally elicited speech errors*. Unpublished doctoral dissertation, University of Toronto.

Dell, G.S. (1985a). Positive feedback in connectionist models: Applications to language production. *Cognitive Science*, *9*, 3–23.

Dell, G.S. (1985b). *Effect of frequency and vocabulary type on phonological speech errors*. Paper presented at the 26th annual meeting of the Psychonomics Society, Boston, MA.

Dell, G.S., & Reich, P.A. (1980). Toward a unified theory of slips of the tongue. In V.A. Fromkin (Ed.), *Errors in linguistic performance: Slips of the tongue, ear, pen and hand* (pp. 273–286). New York: Academic.

Dick, A.O. (1971). Processing time for naming and categorization of letters and numbers. *Perception and Psychophysics*, *9* (38), 350–352.

Duker, S. (1974). Summary of research on time-compressed speech. In S. Duker (Ed.), *Time-compressed speech: An anthology and bibliography in three volumes*. Metuchen, NJ: Scarecrow Press.

Eccles, J.C. (1972). Possible synaptic mechanisms subserving learning. In A.G. Karyman & J.C. Eccles (Eds.), *Brain and human behavior* (pp. 63–92). New York: Springer-Verlag.

Elman, J.L., & McClelland, J.L. (1984). Speech as a cognitive process: The interactive activation model. In N. Lass (Ed.), *Speech and language* (Vol. 10), (pp. 337–374). New York: Academic.

Estes, W.K. (1972). An associative basis for coding and organization in memory. In A.W. Melton & E. Martin (Eds.), *Coding processes in human memory*. Washington, DC: Winston.

Fairbanks, G., & Guttman, N. (1958) Effects of delayed auditory feedback upon articulation. *Journal of Speech and Hearing Research, 1*, 12–22.

Feldman, J.A., & Ballard, D.H. (1982). Connectionist models and their properties. *Cognitive Science, 6*, 205–254.

Fodor, J.A. (1980). *Representations*. Cambridge, MA: MIT Press.

Fodor, J.A. (1983). *The modularity of mind*. Cambridge, MA: MIT Press.

Fodor, J.A., Bever, T.G., & Garrett, M.F. (1974). *The psychology of language*. New York: McGraw-Hill.

Foss, D.J., & Blank, M.A. (1980). Identifying the speech codes. *Cognitive Psychology, 12*, 1–31.

Foss, D.J., & Gernsbacher, M.A. (1983). Cracking the dual code: Toward a unitary model of phoneme identification. *Journal of Verbal Learning & Verbal Behavior, 22*, 609–633.

Fowler, C.A. (1977). *Timing control in speech production*. Bloomington: Indiana University Linguistics Club.

Fowler, C.A., & Tassinary, L.G. (1981). Natural measurement criteria for speech: The anisochrony illusion. In J. Long & A. Baddeley (Eds.), *Attention and performance IX* (pp. 521–535). Hillsdale, NJ: Lawrence Erlbaum.

Foulke, E., & Sticht, T. (1969). Review of research on the intelligibility and comprehension of accelerated speech. *Psychological Bulletin, 72*, 50–62.

Freud, S. (1914). *Psychopathology of everyday life*, A.A. Brill (Trans.). New York: Penguin Books. (Original work published 1901).

Fromkin, V.A. (1973). Appendix. In V.A. Fromkin (Ed.), *Speech errors as linguistic evidence*. Paris: Mouton.

Fromkin, V.A. (1980). Introduction. In V.A. Fromkin (Ed.), *Errors in linguistic performance: Slips of the tongue, ear, pen, and hand* (pp. 1–12). New York: Academic.

Garber, S.R., & Siegel, G.M. (1982). Feedback and motor control in stuttering. In D.L. Routh (Ed.), *Learning, speech, and the complex effects of punishment*. New York: Plenum Press.

Garnes, S., & Bond, Z. (1980). A slip of the ear: A snip of the ear? A slip of the year? In V.A. Fromkin (Ed.), *Errors in linguistic performance: Slips of the tongue, ear, pen and hand* (pp. 231–239). New York: Academic.

Garnham, A., Shillcock, R.C., Brown, G.D.A., Mill, A.I.D., & Cutler, A. (1982). Slips of the tongue in the London-Lund corpus of spontaneous conversation. In A. Cutler (Ed.), *Slips of the tongue, and language production* (pp. 251–263). Amsterdam: Mouton.

Gazdar, G. (1981). Unbounded dependencies and coordinate structure. *Linguistic Inquiry, 12*, 155–184.

Gee, J.P., & Grossjean, F. (1983). Performance structures: A psycholinguistic and linguistic appraisal. *Cognitive Psychology, 15*, 411–458.

Genest, M. (1956). L'Analyse temporelle du travail dactylographique. *Bulletin du Centre d'Etudes et Recherches Psychotechniques, 5*, 183–191.

Gentner, D.R. (1983). Keystroke timing in transcription typing. In W.E. Cooper (Ed.), *Cognitive aspects of skilled typewriting* (pp. 95–120). New York: Springer-Verlag.

Gentner, D.R. (1985). Review of cognition and motor processes. In W. Prinz & A.F. Sanders (Eds.), *Contemporary Psychology, 30* (3), 183–184.

Geschwind, N. (1983). Comments on perceptual processing links. In M. Studdert-Kennedy (Ed.), *Psychobiology of language* (p. 36). Cambridge, MA: MIT Press.

Gleason, H.A. (1961). *Descriptive linguistics*. New York: Holt, Rinehart & Winston.

Gleitman, H., & Jonides, J. (1976). The cost of categorization in visual search: Incomplete processing of targets and field items. *Perception & Psychophysics, 29*(4), 281–288.

Glencross, D.J. (1974). Pauses in a repetitive speed skill. *Perceptual and motor skills, 38,* 246.

Goldstein, M. (1968). Some slips of the tongue. *Psychological Reports, 22,* 1009–1013.

Gordon, P.C., & Meyer, D.E. (1984). Perceptual-motor processing of phonetic features in speech. *Journal of Experimental Psychology: Human Perception and Performance, 10* (1), 153–178.

Gottsdanker, R., & Stelmach, F.E. (1971). The persistence of psychological refractoriness. *Journal of Motor Behavior, 3* (4), 301–312.

Graybiel, A., Kerr, W.A., & Bartley, S.H. (1948). Stimulus thresholds of the semicircular canals as a function of angular acceleration. *American Journal of Psychology, 61,* 21–36.

Greenwald, A.G., & Schulman, H.G. (1973). On doing two things at once: 2. Elimination of the psychological refractory period effect. *Journal of Experimental Psychology, 101,* 70–76.

Grossberg, S. (1982). *Studies of mind and brain: Neural principles of learning, perception, development, cognition and motor control.* Boston: Reidel.

Grudin, J.T. (1981). *The organization of serial order in typing.* Unpublished doctoral dissertation, University of California, San Diego.

Grudin, J.T. (1983). Error patterns in novice and skilled transcription typing. In W.E. Cooper (Ed.), *Cognitive aspects of skilled typewriting* (pp. 121–143). New York: Springer-Verlag.

Hall, J., & Jerger, J. (1978). Central auditory function in stutterers. *Journal of Speech and Hearing Research, 21,* 324–337.

Harley, T.A. (1984). A critique of top-down independent levels models of speech production: Evidence from non-plan-internal speech errors. *Cognitive Science, 8,* 191–219.

Harnad, S. (Ed.), (1986). *Categorical perception.* Cambridge, England: Cambridge University Press.

Hebb, D.O. (1980). *Essay on mind.* Hillsdale, NJ: Erlbaum.

Heffner, R.M.S. (1964). *General phonetics.* Madison, WI: University of Wisconsin Press.

Held, R., & Hein, A. (1963). Movement-produced stimulation in the development of visually guided behavior. *Journal of Comparative and Physiological Psychology, 56,* 872–876.

Henderson, L.A. (1974). A word superiority effect without orthographic assistance. *Quarterly Journal of Experimental Psychology, 26,* 301–311.

Hinrichs, J.V., Mewalt, S.P., & Redding, J. (1973). The Ranschburg effect: Repetition and guessing factors in short-term memory. *Journal of Verbal Learning and Verbal Behavior, 12,* 64–75.

Hinton, G.E., & Sejnowski, T.J. (1986). Learning and relearning in Boltzmann machines. In D.E. Rumelhart, J.L. McClelland, & the PDP Research Group. Parallel distributed processing: *Explorations in the microstructure of cognition, Vol. 1: Foundations.* Cambridge, MA: MIT Press.

Hofstadter, D.R. (1985). *Metamagical themas.* New York: Basic Books.

Hogaboam, T.W., & Perfetti, C.A. (1975). Lexical ambiguity and sentence comprehension. *Journal of Verbal Learning and Verbal Behavior, 14,* 265–274.

Holland, J.H., Holyoak, K., Nisbett, P.R., & Thagard, P. (1986). *Induction: Processes of inference, learning, and discovery.* Cambridge, MA: MIT Press.

Horovitz, L.J., Johnson, S.B., Pearlman, R.C., Schaffer, E.J., & Hedin, A.K. (1978). Stapedial reflex and anxiety in fluent and disfluent speakers. *Journal of Speech and Hearing Research, 21,* 762–767.

Howell, P., & Archer, A. (1984). Susceptibility to the effects of delayed auditory feedback. *Perception & Psychophysics, 36* (3), 296–402.

Howell, P., & Harvey, N. (1983). Perceptual equivalence and motor equivalence in speech. In B. Butterworth (Ed.), *Language production: Vol. 2. Development, writing and other language processes* (pp. 203–224). London: Academic.

Howell, P., & Powell, D.J. (1984). Hearing and voice through bone and air: Implications for explanations of stuttering behavior from studies of normal speakers. *Journal of Fluency Disorders, 9*, 247–264.

Huggins, A.W.F. (1972). On the perception of temporal phenomena in speech. *Journal of the Acoustical Society of America, 51*, 1279–1290.

Hull, F.M. (1952). *Experimental investigation of speech disturbance as a function of the frequency distortion of delayed auditory feedback.* Unpublished doctoral dissertation. University of Illinois.

Huttenlocher, D.P., & Zue, V.W. (1983). Exploring phonotactic and lexical constraints in word recognition. Paper presented to the 106th meeting of the Acoustical Society of America, Nov. 8, San Diego, CA.

Isenberg, D., Walker, C.E.T., Ryder, J.M., & Schweikert, J. (1980, November). *A top-down effect on the identification of function words.* Paper presented at the meeting of the Acoustical Society of America, Los Angeles.

Jahnke, J.C. (1969). The Ranschburg effect. *Psychological Review, 76* (6), 592–605.

James, W. (1890). *The principles of psychology.* New York: Holt.

Jonides, J., & Gleitman, H. (1976). The benefit of categorization in visual search: Target location without identification. *Perception and Psychophysics, 20* (4), 289–298.

Kahneman, D. (1973). *Attention and effort.* Englewood Cliffs, NJ: Prentice-Hall.

Kalmus, H., Benes, P., & Fry, D.B. (1955). Effect of delayed acoustic feedback on some nonvocal activities. *Nature, 175*, 1078.

Keele, S.W. (1973). *Attention and human performance.* Pacific Palisades, CA: Goodyear Publishing.

Keele, S.W. (1981). Behavioral analysis of movement. In V. Brooks (Ed.), *Handbook of physiology: Motor control.* Washington, DC: American Physiological Society.

Keele, S.W. (1986). Motor control. In L. Kaufman, J.P. Thomas, & K. Boff (Eds.), *Handbook of perception and human performance, Vol. 2* (Ch. 30, pp. 1–60). New York: Wiley-Interscience.

Keele, S.W. (1987). Sequencing and timing in skilled perception and action: An overview. In A. Allport, D.G. MacKay, W. Prinz, & E. Scheerer (Eds.), *Language perception and production: Relationships between listening, speaking, reading and writing* (pp. 463–487). London: Academic.

Keele, S.W., & Lyon, D.R. (1982). Individual differences in speech fusion: Methodological and theoretical explorations. *Perception and Psychophysics, 35* (5), 434–442.

Kelso, J.A.S., Southard, D.L., & Goodman, D. (1979). On the coordination of two-handed movements. *Journal of Experimental Psychology: Human Perception and Performance, 5*, 229–238.

Kelso, J.A., & Tuller, B. (1981). Toward a theory of apractic syndromes. *Brain and Language, 12*, 224–245.

Klapp, S.T. (1979). Doing two things at once: The role of temporal compatibility. *Memory and Cognition, 7* (5), 375–381.

Klapp, S.T. (1981). Temporal compatibility in dual motor tasks: II. Simultaneous articulation and hand movements. *Memory and Cognition, 9* (4), 398–401.

Klapp, S.T., Anderson, W.G., & Berrian, R.W. (1973). Implicit speech in reading reconsidered. *Journal of Experimental Psychology, 100*, 368–374.

Klapp, S.T., Hill, M.D., & Tyler, J.G. (1983, November). *Temporal compatibility in dual motor tasks: A perceptual rather than motor interpretation*. Paper presented at the 24th Annual Meeting of the Psychonomic Society, San Diego.

Klapp, S.T., & Wyatt, E.P. (1976). Motor programming within a sequence of responses. *Journal of Motor Behavior, 8* (1), 19–26.

Kodman, E. (1961). Controlled reading rate under delayed speech feedback. *Journal of Auditory Research, 1*, 186–193.

Kolers, P.A. (1968). Bilingualism and information processing. *Scientific American, 218*, 78–85.

Kornhuber, H. (1974). Cerebral cortex, cerebellum, and basal ganglia: An introduction to their motor functions. In F.O. Schmidt & F.G. Worden (Eds.), *The neurosciences* (pp. 333–358). Cambridge, MA: MIT Press.

Kozhevnikov, V.A., & Chistovich, L.A. (1965). *Speech articulation and perception*. Washington, DC: Joint Publications Research Service.

Kuhl, P.K., & Miller, J.D. (1975). Speech perception by the chinchilla: Voice–voiceless distinction in alveolar plosive consonants. *Science, 190*, 69–72.

Kuno, S. (1967). Computer analysis of natural languages. *Symposium on Applied Mathematics, 19*, 52–110.

Lachman, R., Lachman, J.L., & Butterfield, E.C. (1979). *Cognitive psychology and information processing: An introduction*. Hillsdale, NJ: Lawrence Erlbaum.

Lackner, J. (1974). Speech production: Evidence for corollary discharge stabilization of perceptual mechanisms. *Perceptual and Motor Skills, 39*, 899–902.

Lane, H. (1965). Motor theory of speech perception: A critical review. *Psychological Review, 72*, 275–309.

Lashley, K.S. (1951). The problem of serial order in behavior. In L.A. Jeffress (Ed.), *Cerebral mechanisms in behavior* (pp. 112–146). New York: Wiley.

Lassen, N.A., & Larsen, B. (1980). Cortical activity in the left and right hemispheres during language-related brain functions. *Phonetica, 37*, 27–37.

Lecours, A.R. (1966). Serial order in writing. A study of misspelled words in developmental "dysgraphia." *Neuropsychologia, 4*, 221–241.

Lee, B.S. (1950). Effects of delayed speech feedback. *Journal of the Acoustical Society of America, 22*, 824–826.

Lee, D.N., & Lishman, J.R. (1975). Visual proprioceptive control of stance. *Journal of Human Movement Studies, 1*, 87–95.

Leeper, R.A. (1936). A study of a neglected portion of the field of learning: The development of sensory organization. *Journal of Genetic Psychology, 46*, 42–75.

Lehiste, I. (1970). *Suprasegmentals*. Cambridge, MA: MIT Press.

Lehiste, I. (1977). Isochrony reconsidered. *Journal of Phonetics, 5*, 253–263.

Levelt, W.J.M. (1984). Spontaneous self-repairs in speech: Processes and representations. In M.P.R. Van den Broecke & A. Cohen (Eds.), *Proceedings of the Tenth International Congress of Phonetic Sciences* (pp. 105–111). Dordrecht, Holland: Foris.

Levelt, W.J.M., & Cutler, A. (1983). Prosodic marking in speech repair. *Journal of Semantics, 2*, 205–217.

Levelt, W.J.M., Richardson, G., & Heij, W.L. (1985). Pointing and voicing in deictic expressions. *Journal of Memory and Language, 24*, 133–164.

Liberman, A.M., Cooper, F.S., Harris, K.S., & MacNeilage, P.F. (1962). A motor theory of speech perception. In *Proceedings of the speech communication seminar*. Stockholm: Royal Institute of Technology.

Lieberman, P. (1963). Some effects of semantic and grammatical context on the production and perception of speech. *Language and Speech, 6*, 172–187.

Lisker, L. (1978). *Rapid* vs. *rabid*: A catalogue of acoustic features that may cue the distinction. *Haskins Laboratories Status Report on Speech Research*, *54*, 127–132.

MacKay, D.G. (1966). To end ambiguous sentences. *Perception and Psychophysics*, *1*, 426–436.

MacKay, D.G. (1968). Metamorphosis of a critical interval: Age-linked changes in the delay in auditory feedback that produces maximal disruption of speech. *Journal of the Acoustical Society of America*, *19*, 811–821.

MacKay, D.G. (1969a). Effects of ambiguity on stuttering: Towards a theory of speech production at the semantic level. *Kybernetik*, *5*, 195–208.

MacKay, D.G. (1969b). Forward and backward masking in motor systems. *Kybernetik*, *6*, 57–64.

MacKay, D.G. (1969c). The repeated letter effect in the misspellings of dysgraphics and normals. *Perception and Psychophysics*, *5*, 102–106.

MacKay, D.G. (1969d). To speak with an accent: Effects of nasal distortion on stuttering under delayed auditory feedback. *Perception and Psychophysics*, *5*, 183–188.

MacKay, D.G. (1970a). Context-dependent stuttering. *Kybernetik*, *7*, 1–9.

MacKay, D.G. (1970b). How does language familiarity influence stuttering under delayed auditory feedback? *Perceptual and Motor Skills*, *30*, 655–669.

MacKay, D.G. (1970c). Phoneme repetition in the structure of languages. *Language and Speech*, *13* (3), 199–213.

MacKay, D.G. (1970d). Mental diplopia: Towards a model of speech perception at the semantic level. In G. D'Arcais & W.J.M. Levelt (Eds.), *Recent advances in psycholinguistics* (pp. 76–100). Amsterdam: North-Holland.

MacKay, D.G. (1970e). Spoonerisms: The structure of errors in the serial order of speech. *Neuropsychologia*, *8*, 323–350.

MacKay, D.G. (1971). Stress pre-entry in motor systems. *American Journal of Psychology*, *84* (1), 35–51.

MacKay, D.G. (1972). The structure of words and syllables: Evidence from errors in speech. *Cognitive Psychology*, *3*, 210–227.

MacKay, D.G. (1973a). Aspects of the theory of comprehension, memory and attention. *Quarterly Journal of Experimental Psychology*, *25*, 22–40.

MacKay, D.G. (1973b). Complexity in output systems: Evidence from behavioral hybrids. *American Journal of Psychology*, *86* (4), 785–806.

MacKay, D.G. (1974). Aspects of the syntax of behavior: Syllable structure and speech rate. *Quarterly Journal of Experimental Psychology*, *26*, 642–657.

MacKay, D.G. (1978). Speech errors inside the syllable. In A. Bell & J.B. Hooper (Eds.), *Syllables and segments* (pp. 201–212). Amsterdam: North-Holland.

MacKay, D.G. (1979). Lexical insertion, inflection and derivation: Creative processes in word production. *Journal of Psycholinguistic Research*, *8* (5), 477–498.

MacKay, D.G. (1980). Speech errors: Retrospect and prospect. In V.A. Fromkin (Ed.), *Errors in linguistic performance: Slips of the tongue, ear, pen, and hand* (pp. 319–332). New York: Academic.

MacKay, D.G. (1981). The problem of rehearsal or mental practice. *Journal of Motor Behavior*, *13* (4), 274–285.

MacKay, D.G. (1982). The problems of flexibility, fluency, and speed–accuracy trade-off in skilled behavior. *Psychological Review*, *89*, 483–506.

MacKay, D.G. (1983). A theory of the representation and enactment of intentions with applications to the problems of creativity, motor equivalence, speech errors, and automaticity in skilled behavior. In R. Magill (Ed.), *Memory and control of action* (pp. 217–230). Amsterdam: North-Holland.

MacKay, D.G. (1985). A theory of the representation, organization, and timing of action with implications for sequencing disorders. In E.A. Roy (Ed.), *Neuropsychological studies of apraxia and related disorders* (pp. 267–308). Amsterdam: North-Holland.

MacKay, D.G. (1987). Aspects of the theory of action, attention, and awareness. *Perception and Action Research Report*, *120*, Center for Interdisciplinary Research, Bielefeld, FRG, 1–56. Also to appear in W. Prinz & O. Neumann (Eds.), *The functional relations between perception and action*. New York: Springer-Verlag.

MacKay, D.G. (1987). Under what conditions can theoretical psychology survive and prosper? Beyond Greenwald, Pratkanis, Leipe, and Baumgartner (1986) and the empirical epistemology. Accepted for publication, *Psychological Review*.

MacKay, D.G., Allport, A., Prinz, W., & Scheerer, E. (1987). Relationships and modules within language perception–production: An introduction. In A. Allport, D.G. MacKay, W. Prinz, & E. Scheerer (Eds.), *Language perception and production: Relationships between listening, speaking, reading and writing* (pp. 19–22). London: Academic.

MacKay, D.G., & Bever, T.G. (1967). In search of ambiguity. *Perception and Psychophysics*, *2*, 193–200.

MacKay, D.G., & Bowman, R.W. (1969). On producing the meaning in sentences. *American Journal of Psychology*, *82* (1), 23–39.

MacKay, D.G., & Burke, B.D. (1972). Differences between syllable repetition and sentence production under delayed auditory feedback. Unpublished manuscript.

MacKay, D.G., & MacDonald, M. (1984). Stuttering as a sequencing and timing disorder. In W.H. Perkins & R. Curlee (Eds.), *Nature and treatment of stuttering: New directions*. San Diego: College-Hill.

MacKay, D.G., & Soderberg, G. (1971). Homologous intrusions: An analogue of linguistic blends. *Perceptual and Motor Skills*, *32*, 645–646.

MacKay, D.G., & Soderberg, G.A. (1972). Syllable structure and long range interactions in the production of speech. Unpublished manuscript.

MacKay, D.M. (1973). Visual stability and voluntary eye movements. In H. Autrum, R. Jung, W.R. Loewnstein, D.M. MacKay, & H.-L. Teuber (Eds.), *Handbook of sensory physiology* (pp. 307–331). New York: Springer-Verlag.

MacKay, D.M. (1984). Evaluation: The missing link between cognition and action. In W. Prinz & A. Saunders (Eds.), *Cognition and motor processes* (pp. 175–184). West Berlin: Springer-Verlag.

Mandler, G. (1985). *Cognitive psychology: An essay in cognitive science*. Hillsdale, NJ: Lawrence Erlbaum.

Marr, D., & Poggio, T. (1977). From understanding computation to understanding neural circuitry. *Neurosciences Research Progress Bulletin*, *15*, 470–488.

Marslen-Wilson, W.D. (1975). Sentence perception as an interactive parallel process. *Science*, *189*, 226–228.

Marslen-Wilson, W.D., & Tyler, L.K. (1980). The temporal structure of spoken language understanding. *Cognition*, *8*, 1–71.

Marslen-Wilson, W.D., & Welsh, A. (1978). Processing interactions and lexical access during word recognition in continuous speech. *Cognitive Psychology*, *10*, 29–63.

Martin, J.H. (1985). Cortical neurons, the EEG, and the mechanisms of epilepsy. In E.R. Kandel & J.H. Schwartz (Eds.), *Principles of neural science*. Amsterdam: Elsevier.

Massaro, D.W. (1979). Reading and listening. In P.A. Kolers, M.E. Wrolstad, & H. Bouma (Eds.), *Processing of visible language* (pp. 331–354). New York: Plenum Press.

Massaro, D.W. (1981). Sound to representation: An information processing analysis. In T. Meyers, J. Laver, & J. Anderson (Eds.), *The cognitive representation of speech* (pp. 181–193). New York: Academic.

Massaro, D.W., & Cohen, M.M. (1976). The contribution of fundamental frequency and voice onset time to the /zi/–/si/ distinction. *Journal of the Acoustical Society of America, 60*, 704–717.

Massaro, D.W., & Cohen, M.M. (1983a). Categorical or continuous speech perception: A new test. *Speech Communication, 2*, 15–35.

Massaro, D.W., & Cohen, M.M. (1983b). Phonological constraints in speech perception. *Perception & Psychophysics, 34*, 338–348.

Mateer, C.A. (1983). Motor and perceptual functions of the left hemisphere and their interaction. In S.J. Segalowitz (Ed.), *Language functions and brain organization* (pp. 145–170). New York: Academic.

McClelland, J.L., & Elman, J.L. (1986). The TRACE model of speech perception. *Cognitive Psychology, 18*, 1–86.

McClelland, J.L., & Kawamoto, A.H. (1986). Mechanisms of sentence processing: Assigning roles to constituents of sentences. In McClelland, J.L., Rumelhart, D.E., & the PDP Research Group (Eds.), *Parallel distributed processing. Explorations in the microstructure of cognition. Volume 2: Psychological and Biological Models* (pp. 272–325). Cambridge, MA: MIT Press.

McClelland, J.L., & Rumelhart, D.E. (1981). An interactive–activation model of context effects in letter perception: 1. An account of basic findings. *Psychological Review, 88*, 375–407.

McClelland, J.L., Rumelhart, D.E., & the PDP Research Group. (1986). *Parallel distributed processing. Explorations in the microstructure of cognition: Vol. 2. Psychological and Biological Models*. Cambridge, MA: MIT Press.

McFarlane, S.J., & Prins, D. (1978). Neural response time of stutterers and nonstutterers. *Journal of Speech and Hearing Research, 21*, 768–778.

McGurk, H., & MacDonald, J. (1976). Hearing lips and seeing voices. *Nature, 264*, 746–748.

McLeod, P., & Posner, M.I. (1983). Privileged loops from percept to act. In H. Bouma & D.G. Bouwhuis (Eds.), *Attention and performance X: Control of language processes* (pp. 55–69). Hillsdale, NJ: Lawrence Erlbaum.

Meringer, R. (1908). *Aus dem Leben der Sprache*. Berlin: Behrs Verlag.

Meringer, R., & Mayer, K. (1895). *Versprechen und Verlesen*. Stuttgart: Goeschensche Verlagsbuchhandlung.

Meyer, D.E., & Gordon, P.C. (1983). Dependencies between rapid speech perception and production: Evidence for a shared sensory–motor voicing mechanism. In H. Bouma & D. Bouwhuis (Eds.), *Attention and performance X, Control of language processes*. Hillsdale, NJ: Lawrence Erlbaum.

Meyer, D.E., Smith, E.K., & Wright, C.E. (1982). Models for the speed and accuracy of aimed movements. *Psychological Review, 89*, 483–506.

Michaels, C.F., & Carello, C. (1981). *Direct perception*. Englewood Cliffs, NJ: Prentice-Hall.

Miller, G.A., Gallanter, E., & Pribram, K.H. (1960). *Plans and the structure of behavior*. New York: Holt, Rinehart & Winston.

Mittelstaedt, H. (1985). *The information processing structure of the subjective vertical: A cybernetic bridge between its psychophysics and its neurobiology* (Report No. 60). Research Group on Perception and Action at the Center for Interdisciplinary Research (ZiF). West Germany: University of Bielefeld.

Morton, J. (1969). Interaction of information in word recognition. *Psychological Review, 76*, 165–178.

Neisser, U. (1976). *Cognition and reality.* San Francisco: W. H. Freeman.

Neisser, U. (1982). Understanding psychological man. *Psychology Today, 16,* 40.

Neumann, O. (1984). *Delayed auditory feedback and the control of speech production* (Report No. 6). Research Group on Perception and Action at the Center for Interdisciplinary Research. West Germany: University of Bielefeld.

Newell, A. (1973). You can't play 20 questions with nature and win: Projective comments on the papers of this symposium. In W.E. Chase (Ed.), *Visual information processing* (pp. 283–308). New York: Academic.

Nickerson, R. (1973). Can characters be classified directly as digits vs. letters or must they be identified first? *Memory & Cognition, 1* (4), 477–484.

Nooteboom, S.G. (1980). Speaking and unspeaking: Detection and correction of phonological and lexical errors in spontaneous speech. In V.A. Fromkin (Ed.), *Errors in linguistic performance. Slips of the tongue, ear, pen and hand* (pp. 87–95). New York: Academic.

Norman, D.A. (1969). Memory while shadowing. *Quarterly Journal of Experimental Psychology, 21,* 85–93.

Norman, D.A. (1981). Categorization of action slips. *Psychological Review, 88* (1), 1–15.

Norman, D.A., & Rumelhart, D.E. (1983). Studies of typing from the LNR research group. In W.E. Cooper (Ed.), *Cognitive aspects of skilled typewriting* (pp. 45–66). New York: Springer-Verlag.

Oatley, K. (1978). *Perceptions and representations.* London: Methuen.

Ohala, J., & Hirano, M. (1967). *Control mechanisms for the sequencing of neuromuscular events in speech.* Paper presented at the meeting of the AFCRL, MIT, Cambridge, MA.

Ojemann, G.A. (1983). Brain organization for language from the perspective of electrical stimulation mapping. *The Behavioral and Brain Sciences, 2,* 189–230.

Perkins, W.H., Bell, J., Johnson, L., & Stocks, J. (1979). Phone rate and the effective planning time hypothesis of stuttering. *Journal of Speech and Hearing Research, 22,* 747–755.

Pick, H.L., & Saltzman, E. (1978). Perception of communicative information. In H.L. Pick & E. Saltzman (Eds.), *Modes of perceiving and processing information* (pp. 1–20). Hillsdale, NJ: Lawrence Erlbaum.

Pisoni, D.B. (1975). Auditory short-term memory and vowel perception. *Memory and Cognition, 3,* 17–18.

Pisoni, D.B., & Tash, J. (1974). Reaction times to comparisons within and across phonetic categories. *Perception and Psychophysics, 15, 2,* 285–290.

Pokorny, R.A. (1985). *Searching for interaction between timing of motor tasks and timing of perceptual tasks.* Unpublished doctoral dissertation, University of Oregon, Eugene, OR.

Porter, R.J., & Lubker, J.F. (1980). Rapid reproduction of vowel–vowel sequences: Evidence for a fast and direct acoustic–motor linkage in speech. *Journal of Speech and Hearing Research, 23,* 593–602.

Posner, M.I. (1978). *Chronometric explorations of mind.* Hillsdale, NJ: Lawrence Erlbaum.

Preus, A. (1981). *Identifying subgroups of stutterers.* Oslo: University of Oslo Press.

Pribram, K.H. (1971). *Languages of the brain: Experimental paradoxes and principles in neuropsychology.* Englewood Cliffs, NJ: Prentice-Hall.

Prinz, W. (1985). *Continuous selection* (Report No. 74). Research Group on Perception and Action at the Center for Interdisciplinary Research (ZiF). West Germany: University of Bielefeld.

Prinz, W., Meinecke, C., & Hielscher, M. (1985). *Effect of stimulus degradation on category search* (Report No. 27). West Germany: University of Bielefeld (ZiF).

Prinz, W., & Nattkemper, D. (1985). *Effects of secondary tasks on search performance* (Report No. 36). Research Group on Perception and Action at the Center for Interdisciplinary Research (ZiF). West Germany: University of Bielefeld.

Rabbitt, P.M.A., & Phillips, S. (1967). Error-detection and correction latencies as a function of S-R compatibility. *Quarterly Journal of Experimental Psychology, 19*, 37–42.

Rabbitt, P.M.A., Vyas, S.M., & Fearnley, S. (1975). Programming sequences of complex responses. In P.M.A. Rabbitt & S. Dornic (Eds.), *Attention and performance V* (pp. 395–417). London: Academic.

Ranschburg, P. (1902). Ueber Hemmung gleichzeitiger Reizwirkungen. *Zeitschrift für Psychologie, 30*, 39–86.

Ratner, G.S., Gawronski, J.J., & Rice, F.E. (1964). The variable of concurrent action in the language of children: Effects of delayed speech feedback. *Psychological Reports, 14*, 47–56.

Reich, A., Till, J., & Goldsmith, H. (1981). Laryngeal and manual reaction times of stuttering and nonstuttering adults. *Journal of Speech and Hearing Research, 24*, 192–196.

Repp, B. (1982). Phonetic trading relations and context effects: New experimental evidence for a speech mode of perception. *Psychological Bulletin, 92* (1), 81–110.

Roehrig, W.C. (1965). Addition of controlled distortion to delay of auditory feedback. *Perceptual and Motor Skills, 21*, 407–413.

Rosenbaum, D.A. (1985). Motor programming: A review and scheduling theory. In H. Heuer, U. Kleinbeck, & K.-H. Schmidt (Eds.), *Motor behavior: Programming, control, and acquisition* (pp. 1–33). West Berlin: Springer-Verlag.

Rosenblueth, A., Wiener, N., & Bigelow, J. (1943). Behavior, purpose and teleology. *Philosophy of Science, 10* (1), 18–24.

Rubinstein, L. (1964). Intersensory and intrasensory effects in simple reaction time. *Perceptual and Motor Skills, 18*, 159–172.

Rumelhart, D.E., McClelland, J.L., & Hinton, G.E. (1986). The appeal of parallel distributed processing. In. D.E. Rumelhart, J.L. McClelland, & the PDP Research Group (Eds.), *Parallel distributed processing. Explorations in the microstructure of condition: Vol. 1. Foundations* (pp. 3–44). Cambridge, MA: MIT Press.

Rumelhart, D.E., McClelland, J.L., & the PDP Research Group (1986). *Parallel distributed processing. Explorations in the microstructure of cognition: Vol. 1. Foundations.* Cambridge, MA: MIT Press.

Rumelhart, D.E., Smolensky, P., McClelland, J.L., & Hinton, G.E. (1986). Schemata and sequential thought processes in PDP models. In J.L. McClelland, D.E. Rumelhart, & the PDP Research Group (Eds.), *Parallel distributed processing. Explorations in the microstructure of cognition: Vol. 2. Psychological and biological models* (pp. 7–57). Cambridge, MA: MIT Press.

Salasoo, A., & Pisoni, D. (1985). Interaction of knowledge sources in spoken word identification. *Journal of Memory and Language, 24*, 210–231.

Samuel, A.G. (1981). The role of bottom-up confirmation in the phonemic restoration illusion. *Journal of Experimental Psychology: Human Perception and Performance, 7* (5), 1124–1131.

Savin, H.B., & Bever, T.G. (1970). The nonperceptual reality of the phoneme. *Journal of Verbal Learning and Verbal Behavior, 9*, 295–302.

Schmidt, R.A. (1980). On the theoretical status of time in motor-program representations. In G.E. Stelmach & J. Requin (Eds.), *Tutorials in motor behavior.* Amsterdam: North-Holland.

Schmidt, R.A. (1982). *Motor control and learning: A behavioral emphasis*. Champaign, IL: Human Kinetics.

Sejnowski, T.J. (1981). Skeleton filters in the brain. In G.E. Hinton & J.A. Anderson (Eds.), *Parallel models of associative memory* (pp. 49–82). Hillsdale, NJ: Lawrence Erlbaum.

Seo, H. (1974). Computer speech compression. In S. Duker (Ed.), *Time-compressed speech: An anthology and bibliography in three volumes*. Metuchen, NJ: Scarecrow Press.

Seymour, W.D. (1959). Experiments on the acquisition of industrial skills: Part 4. Assembly tasks. *Occupational Psychology, 33*, 18–35.

Shaffer, L.H. (1978). Timing in the motor programming of typing. *Quarterly Journal of Experimental Psychology, 30*, 333–345.

Shaffer, L.H. (1980). Analysing piano performance. In G.E. Stelmach & J. Requin (Eds.), *Tutorials in motor behavior* (pp. 443–455). Amsterdam: North-Holland.

Shaffer, L.H. (1982). Rhythm and timing in skill. *Psychological Review, 89* (2), 109–122.

Shapiro, D.C. (1977). A preliminary attempt to determine the duration of a motor program. In D.M. Landers & R.W. Christina (Eds.), *Psychology of motor behavior and sports*. Champaign, IL: Human Kinetics.

Shapiro, D.C., Zernicke, R.F., Gregor, R.J., & Diestel, J.D. (1981). Evidence for generalized motor programs using gait pattern analysis. *Journal of Motor Behavior, 13*, 33–47.

Shearer, W., & Simmons, F. (1965). Middle ear activity during speech in normal speakers and stutterers. *Journal of Speech and Hearing Research, 8*, 203–237.

Shields, J.L., McHugh, A., & Martin, J.G. (1974). Reaction time to phoneme targets as a function of rhythmic cues in continuous speech. *Journal of Experimental Psychology, 102*, 250–255.

Shiffrin, R.M., & Schneider, W. (1977). Controlled and automatic human information processing: 2. Perceptual learning, automatic attending, and a general theory. *Psychological Review, 84*, 127–190.

Siegel, G.M., & Pick, H.L. (1974). Auditory feedback in the regulation of voice. *Journal of the Acoustical Society of America, 56*, 1618–1624.

Simon, H.A. (1980). How to win at twenty questions with nature. In R.A. Cole (Ed.), *Perception and production of fluent speech* (pp. 535–548). Hillsdale, NJ: Lawrence Erlbaum.

Simmons, F.B. (1964). Perceptual theories of middle ear function. *Annals of Oto-rhino-laryngology, 73*, 73.

Smith, K.U. (1962). *Delayed sensory feedback and behavior*. Philadelphia: W. B. Saunders.

Spoer, K.T., & Smith, E.E. (1973). The role of syllables in perceptual processing. *Cognitive Psychology, 5*, 71–89.

Steinbach, M. (1985, February). *Behavioral and anatomical evidence for eye muscle proprioception in surgically-treated humans*. Paper presented at the conference on sensorimotor interaction in space perception and action at the Center for Interdisciplinary Research (ZiF), University of Bielefeld, West Germany.

Stemberger, J.P. (1983). The nature of /r/ and /l/ in English: Evidence from speech errors. *Journal of Phonetics, 11*, 139–147.

Stemberger, J.P. (1985). An interactive activation model of language production. In A. Ellis (Ed.), *Progress in the psychology of language* (pp. 143–186). London: Lawrence Erlbaum.

Sternberg, S., Monsell, S., Knoll, R.L., & Wright, C.E. (1978). The latency and duration of rapid movement sequences: Comparisons of speech and typewriting. In G.E. Stelmach (Ed.), *Information processing in motor control and learning*. New York: Academic.

Straight, S. (1980). Auditory vs. articulatory phonological processes and their development in children. In G.H. Yeni-Komshian, J.F. Kavanagh, & C.A. Ferguson (Eds.), *Child phonology: Vol. 1. Production*. New York: Academic.

Studdert-Kennedy, M. (1983). Perceptual processing links to the motor system. In M. Studdert-Kennedy (Ed.), *Psychobiology of language* (pp. 29–39). Cambridge, MA: MIT Press.

Summerfield, Q., & Haggard, M. (1977). On the dissociation of spatial and temporal cues to the voicing distinction in initial stop consonants. *Journal of the Acoustical Society of America*, *62*, 435–438.

Swinney, D.A. (1979). Lexical access during sentence comprehension: (Re)consideration of context effects. *Journal of Verbal Learning and Verbal Behavior*, *18*, 645–659.

Tent, J., & Clark, J.E. (1980). An experimental investigation into the perception of slips of the tongue. *Journal of Phonetics*, *8*, 317–325.

Thagard, P., & Holyoak, K. (1985). Discovering the wave theory of sound: Induction in the context of problem solving. In *Proceedings of the Ninth International Joint Conference on Artificial Intelligence*. Palo Alto: William Kaufmann.

Thorndike, E.L. (1898). Animal intelligence: An experimental study of the associative processes in animals. *Psychological Monographs*, *1* (4, Whole No. 8).

Treiman, R. (1983). The structure of spoken syllables: Evidence from novel word games. *Cognition*, *15*, 49–74.

Tuller, B., Kelso, J.A.S., & Harris, K.S. (1982). Interarticulator phasing as an index of temporal regularity in speech. *Journal of Experimental Psychology: Human Perception and Performance*, *8* (3), 460–472.

Turvey, M.T. (1977). Preliminaries to a theory of action with reference to vision. In R.E. Shaw & J. Bransford (Eds.), *Perceiving, acting and knowing*. Hillsdale, NJ: Lawrence Erlbaum.

Turvey, M.T., Shaw, R.E., Reed, E.S., & Mace, W.M. (1981). Ecological laws of perceiving and acting: In reply to Fodor and Pylyshyn (1981). *Cognition*, *9*, 237–304.

Van Lanker, D., & Carter, G.J. (1981). Idiomatic vs. literal interpretations of ditropically ambiguous sentences. *Journal of Speech and Hearing Research*, *24*, 64–69.

Van Riper, C. (1982). *The nature of stuttering*. Englewood Cliffs, NJ: Prentice-Hall.

von Bekesy, G. (1967). *Sensory inhibition*. Princeton, NJ: Princeton University Press.

von Treba, P., & Smith, K.U. (1952). The dimensional analysis of motion: 4. Transfer effects and direction of movement. *Journal of Applied Psychology*, *36*, 348–353.

Warren, R.M. (1968). Verbal transformation effect and auditory perceptual mechanisms. *Psychological Bulletin*, *70*, 261–270.

Warren, R.M. (1970). Perceptual restoration of missing speech sounds. *Science*, *167*, 392–393.

Warren, R.M. (1974). Auditory temporal discrimination by trained listeners. *Cognitive Psychology*, *6*, 237–256.

Warren, R.M. (1982). *Auditory perception: A new synthesis*. New York: Pergamon Press.

Warren, R.M., & Sherman, G.L. (1974). Phonemic restorations based on subsequent context. *Perception and Psychophysics*, *16*, 150–156.

Warren, R.M., & Warren, R.P. (1970). Auditory illusions and confusions. *Scientific American*, *223*, 30–36.

Wehrkamp, R., & Smith, K.U. (1952). Dimensional analysis of motion: 2. Travel–distance effects. *Journal of Applied Psychology*, *36*, 201–206.

Welford, A.T. (1968). *Fundamentals of skill*. London: Methuen.

Wickelgren, W. (1979). *Cognitive psychology*. Englewood Cliffs, NJ: Prentice-Hall.

Wing, A.M. (1978). Response timing in handwriting. In G.E. Stelmach (Ed.), *Information processing in motor control and learning* (pp. 469–485). New York: Academic.

Witkin, H.A., Wapner, S., & Leventhall, T. (1952). Sound localization with conflicting visual and auditory cues. *Journal of Experimental Psychology, 43*, 57–58.

Woodworth, R.S. (1938). *Experimental psychology.* New York: Holt.

Wyneken, C. (1868). Ueber das Stottern und dessen Heilung. *Zeitschuift fur nat. Med.*, 1–29.

Zimmermann, G.N. (1980). Articulatory behavior associated with stuttering: A cineradiographic analysis. *Journal of Speech and Hearing Disorders, 23*, 108–121.

Index of Names

Subject Index